Y0-BQE-471

Greening Brownfields: Remediation Through Sustainable Development

ABOUT THE AUTHOR

William Sarni is founder and CEO of DOMANI, a consulting firm that provides innovative business and technical sustainability solutions to companies committed to increasing revenue, mitigating risk, and improving operating efficiency. He has managed a wide range of complex sustainability and environmental programs throughout the United States, Europe, and Asia. Mr. Sarni is a member of the Environmental Compliance Committee of the Chicago Climate Exchange and is active with the Conference Board and Sustainable Life Media/Sustainable Brands.

Greening Brownfields: Remediation Through Sustainable Development

William Sarni

NEW HANOVER COUNTY
PUBLIC LIBRARY
201 CHESTNUT STREET
WILMINGTON, NC 28401

Mc
Graw
Hill

New York Chicago San Francisco Lisbon London Madrid
Mexico City Milan New Delhi San Juan Seoul
Singapore Sydney Toronto

The *McGraw-Hill* Companies

Cataloging-in-Publication Data is on file with the Library of Congress

Copyright © 2010 by The McGraw-Hill Companies, Inc. All rights reserved. Printed in the United States of America. Except as permitted under the United States Copyright Act of 1976, no part of this publication may be reproduced or distributed in any form or by any means, or stored in a data base or retrieval system, without the prior written permission of the publisher.

1 2 3 4 5 6 7 8 9 0 DOC/DOC 0 1 5 4 3 2 1 0 9

ISBN 978-0-07-160909-8
MHID 0-07-160909-1

Sponsoring Editor
Joy Bramble Oehlkers

Production Supervisor
Pamela A. Pelton

Editing Supervisor
Stephen M. Smith

Acquisitions Coordinator
Michael Mulcahy

Project Manager
Jacquie Wallace, Lone Wolf Enterprises, Ltd.

Copy Editor
Jacquie Wallace, Lone Wolf Enterprises, Ltd.

Proofreader
Leona Woodson, Lone Wolf Enterprises, Ltd.

Art Director, Cover
Jeff Weeks

Composition
Lone Wolf Enterprises, Ltd.

Printed and bound by RRDonnelley.

McGraw-Hill books are available at special quantity discounts to use as premiums and sales promotions, or for use in corporate training programs. To contact a representative, please e-mail us at bulksales@mcgraw-hill.com.

 The pages within this book were printed on acid-free paper containing 100% postconsumer fiber.

Information contained in this work has been obtained by The McGraw-Hill Companies, Inc. ("McGraw-Hill") from sources believed to be reliable. However, neither McGraw-Hill nor its authors guarantee the accuracy or completeness of any information published herein, and neither McGraw-Hill nor its authors shall be responsible for any errors, omissions, or damages arising out of use of this information. This work is published with the understanding that McGraw-Hill and its authors are supplying information but are not attempting to render engineering or other professional services. If such services are required, the assistance of an appropriate professional should be sought.

This book is dedicated to my family, Maureen, James, Thomas, and Charles,
who share a hopeful vision for the future and are committed to shaping it;
to my parents, Josephine Napoli and Michael Sarni;
and to my sister, Celeste Sarni.

CONTENTS

CHAPTER 4: INCENTIVES   55

CHAPTER 5: FRAMEWORKS FOR GREENING BROWNFIELDS   77

CHAPTER 6: LAND PLANNING   91

CHAPTER 7: GREEN BUILDING   127

CHAPTER 8: VALUING FINANCIAL RETURN OR BEYOND ASSET CONVERSION   149

CHAPTER 9: CASE STUDIES   177

FOREWORD

A powerful green wave is moving through the business world and society at large—a rising interest in and concern about the environment. The pressures on businesses and communities to go "green" are evolving and growing, in good economic times and bad. Corporate executives are facing new resource constraints, rising regulations, mega-forces like technology and globalization changing how the world works, and questions from a wide range of stakeholders about their environmental and social performance.

Driving many of these pressures is a set of very real and growing environmental challenges, including the global climate crisis, threats to water quality and quantity, increasing fears about chemicals and toxics, and concerns about loss of biodiversity and how we use land. To tackle these issues, governments at the local, state, national, and international levels are all forcing companies to operate with higher environmental standards. And of course international negotiations about how to tackle the largest environmental problem, climate change, continue to evolve. One near-certain outcome is a price on carbon, which will drastically change everything from how we power our world to how we get around.

The mega-forces also include population growth and the rise of middle-class consumers—some one billion new ones over the coming generation—who are demanding more of everything. This pressure will continue to make natural resources, including land, harder to come by and more expensive. Rising calls for transparency are also forcing companies and communities to disclose more about what's in every product, how it's made, and what happens to it at the end of its life.

Finally, stakeholders are focusing all these pressures onto companies in new ways. As part of a "greening of the supply chain" movement, business customers are asking their own suppliers to reduce energy, water, and resource use. They are also measuring and comparing suppliers on carbon footprints and product toxicity, among other things. The end customers, consumers, are looking for better social and environmental performance from the companies they buy from, and employees want more meaning and higher standards from the companies they work for.

In total, this melting pot of pressures is forcing significant change on communities, companies, and even individuals. But luckily for all of us, the change is good. Greener businesses are more profitable and sustainable enterprises, and greener communities are healthier and more attractive to live in.

At its core, going "green" means doing more with less—finding a way to satisfy our needs and create a quality of life with dramatically lower energy, water, and other resource use. A big part of this strategy and philosophy is to eliminate the concept of waste. To reuse something that seems like it has no value and create something new from it lies at the core of sustainability. And this is where brownfields come in. Those pieces of land lying fallow, many of which are literally contaminated, represent tremendous opportunities for renewal. Taking something that's going to waste and making it useful and beautiful again is perhaps the highest act of sustainability. The greening of brownfields is literally repairing the earth. What could be greener than that?

But aside from answering a higher calling, the greening of brownfields represents a tangible opportunity for businesses in a range of sectors. Construction companies that understand green practices will get more business. Real estate developers can create commercial or residential buildings that will be in great demand as buyers look for more sustainable options. The evidence is mounting that people will pay more for greener space. In recent years, commercial buildings that are "LEED" certified (the green building scoring system prominent in the United States) have commanded higher rents and higher occupancy rates.

But it's not just the private sector that benefits. At the municipal level, reusing space sitting in the heart of communities reduces sprawl, eliminates fights about using virgin land, shortens commutes, and raises real estate values and tax revenues. The greening of brownfields also creates a big opportunity to develop and implement new building technologies. Since buildings represent over 40 percent of all energy use and greenhouse gas emissions, new designs are imperative if we want to solve our energy and climate problems. And a community renewal initiative that pursues green principles will help our most blighted areas leapfrog to become the cleanest and most innovative around.

At the global level, consider this: More than half of humanity now lives in cities. Where will they all live and work? How can we keep the environmental impacts of rising demands and consumption to a minimum?

We all face a pressing need to provide a quality of life for what will be nine billion people, and do so using drastically less stuff. We need to bring carbon emissions down 80 percent by mid-century to avoid what scientists tell us will be the most dangerous impacts of climate change. Incorporating the growth of population into already dense areas, reusing land, and creating buildings in completely new ways that are low impact and even carbon-neutral—these are the building blocks for solving our biggest problems. The greening of brownfields will be one of the critical paths to sustainability at all levels—financial, environmental, and social. Greening brownfields is an important contribution to a vital movement.

Andrew Winston
Author, *Green Recovery,* and co-author, *Green to Gold*

ACKNOWLEDGMENTS

There is no shortage of people to acknowledge who provided inspiration, encouragement, support, and their time for this effort.

First, I would like to thank my wife, Maureen Meegan, who provided constant encouragement and gave up countless hours with me while I worked on the book; and also my sons, James, Thomas, and Charles, who always encouraged me to "finish the book!".

I would also like to thank the following people:

My extended Sarni, Meegan, Casey, and Zelkovich families and, in particular, Mike Casey and Mary Helen Zelkovich, who pushed me along through the process.

Ljubomir Domijan and Stokijier for their guidance.

Ben Dukes, Frank Sherman, Deanna "Drai" Turner, Mark Miller, and Matthew Segur for helping with the research and graphics for the book in their spare time.

Andrew Winston for writing the foreword for the book, providing constant support in moving the sustainability agenda forward, and encouraging me to "keep the faith."

Paul Roux, chairman of the board of Roux Associates, for providing me with my first job opportunity and teaching me about hydrogeology and brownfields remediation; and Doug Swanson, the CEO of Roux Associates, for his support and being a friend when one was needed.

Clients such as Enterprise Community Partners, Cherokee Investment Partners, and Invensys, who had the vision to transform brownfield sites into green development projects.

R. Dodge Woodson and Jacquie Wallace for guiding me through the process of writing, and Joy Bramble Oehlkers of McGraw-Hill for supporting the publishing of the book.

Finally, all the innovators and visionaries who provided the inspiration to believe in sustainability and how it will transform business and civil society.

William Sarni

1

Global Trends—
The Perfect Storm

What are the trends that are driving the increased development of brownfield sites and especially the green development of brownfield sites? It is no coincidence that there are several business and public policy trends that are driving the greening of brownfields.

The perfect storm of business opportunity and public policy is driving brownfield redevelopment from mere liability management to high value green development projects. The reasons for increased brownfield development are multifold: there are fewer "clean" undeveloped urban properties; remediation technologies are becoming cheaper and more predictable (in terms of achieving clean up goals within a specified time frame); and state and federal incentives are in place to encourage redevelopment.

WHY BROWNFIELD SITES?

Let's briefly examine why brownfield sites are appealing as development projects in spite of greater complexity and longer development time. In general, brownfield sites can be cheaper overall to develop and are located in highly desirable urban infill areas.

An excellent example is the Atlantic Station site in Atlanta, Georgia. This site is a 138-acre, 12 million square foot mixed-use project at the former Atlantic Steel Mill site.[1] The land was acquired by Jacoby Development Corporation in 1999 for approximately $76 million (U.S.). The remediation costs for the site were approximately $25 million (U.S.) which results in a per acre cost of $731,000. This compares to the nearby development site for the Atlanta Symphony which cost approximately $22.3 million (U.S.) for 6.36 acres or $3.5 million per acre. Not a bad deal for the Atlantic Station property in comparison.

In addition to favorable economics, attitudes toward the development of brownfield sites have changed. For example, in 2002 the U.S. federal government passed the Small Business Liability Relief and Brownfields Revitalization Act that provided greater definition of potential purchaser liability under the 1980 Comprehensive Environmental Response Compensation and Liability Act (CERCLA). CERCLA was also previously amended in 1986 adding the "innocent landowner" defense which provided an exemption for property owners if they performed due diligence to identify any site pollution. With increased liability definition, developers had an increased ability to quantify potential remediation costs and manage any potential liability.

The availability of insurance coverage to manage liability and unexpected costs from remediation and risk based cleanup goals (instead of pre-established cleanup goals) provides greater economic certainty for developers taking on brownfield sites.

Key insurance products available for developers are remediation cost cap insurance and environmental liability insurance. Briefly, cost cap insurance provides coverage for developers where contaminants are known. Environmental liability insurance covers costs due to the discovery of contaminants unknown at the time the remediation program was developed as well as provides coverage for changes in regulations and third party liability.

The real sign that the development of brownfield sites is becoming commonplace is the presence of mainstream developers in the market. For example, Lennar Corporation and LNR are planning to develop 680 acres of land on Mare Island in San Francisco Bay. The plan is to build 1500 homes and 6 million square feet of commercial real estate.

Although we will discuss selected case studies in detail later, it is worth highlighting landmark brownfield projects to provide a feel for the impact brownfield redevelopment has had on urban areas over the past several years.

Most notable is the transformation of Pittsburgh's steel mills into vibrant urban development projects. The decline of the U.S. steel industry in the 1970s and 1980s provided Pittsburgh with the "opportunity" to redevelop thousands of acres of riverfront property. The credit for successful redevelopment of Pittsburgh goes in part to The Urban Redevelopment Authority of Pittsburgh (URAP). URAP worked with developers, local and state governments, and foundations to deal with contaminated properties and transform large tracts of land into usable urban projects.

These development projects include Washington's Landing at Herr's Island (Washington Landing), a 42-acre site on the western bank of the Allegheny River. Abandoned stockyards and salvage operations were remediated at a cost of approximately $2.5 million, creating a mixed-use development project which includes commercial, manufacturing, recreation, and housing. An added benefit was the creation of at least 2100 new jobs and reportedly generating over $1 million in revenue for the City of Pittsburgh.

Other Pittsburgh redevelopment projects include the Pittsburgh Technology Center, the South Side Works, and Summerset at Frick Park, all outstanding examples of successful brownfield redevelopment projects.

So, why would brownfield sites be built green? Let's briefly examine the rationale for greening brownfield sites.

WHY GREENING BROWNFIELDS?

The bottom line is that while the redevelopment of brownfield sites creates value from a liability ("asset conversion"), much greater value can be created if the site incorporates sustainable land use and green building practices. Employing sustainable practices creates greater brand value (especially for a corporation redeveloping a brownfield site) and vastly improves operating efficiency over the life cycle of the development.

The Ford Motor Company has been proactive in the redevelopment of brownfield sites, and an excellent example is its Fairlane Green site in Allen Park, Michigan.

The Fairlane Green project is an approximately 243-acre retail and recreational center built at the remediated Ford Allen Park

Clay Mine Landfill. The project is also significant as the real estate unit of Ford (Ford Land) worked to sustainably redevelop the site, not merely the development of the landfill. The greening of the site is central to the redevelopment plan. Two-thirds of the site will be open space, including fields, ponds, trails, and a planned 43-acre park surrounding an estimated 1 million square feet of shops and restaurants.

In addition to sustainable land use practices, phase 1 of the project is the first multi-tenant retail development to achieve gold-level Leadership in Energy and Environmental Design (LEED) certification from the United States Green Building Council (USGBC). The LEED Gold development includes high efficiency, CFC-free heating and cooling equipment, reflective roofing, low-emitting materials, water-efficient plumbing fixtures, recycled and locally sourced building materials, windows and skylights, and a cistern to capture and reuse rainwater.

Moreover, the Fairlane Green "Brown to Green"™ development includes several innovations:

- Currently the largest retail development in the U.S. built on a landfill and the largest landfill redevelopment in the state of Michigan.

- The first development in Michigan to use a three-dimensional legal description to separate the landfill from the surface development. This allows Ford to retain landfill ownership and responsibility while selling the surface to owners and developers.

- In 2002, it received what was at the time the largest tax increment financing (TIF) package offered by the state. The TIF covers brownfield-related development costs—measures to reduce settlement, protect the landfill cap, reinforce slopes, and construct utilities.

- Fairlane Green is one of the first vertical construction projects built on Styrofoam-like blocks called geofoam. Geofoam's light weight reduces the potential for future settlement.

The redevelopment of the Fairlane Green brownfield site won Ford the Phoenix Award for excellence in the successful development of a brownfield site (www.phoenixawards.org) presented at the National Brownfields Conference in May 2008. The Fairlane Green site included the redevelopment of an old landfill site into "a sustainable new development that provides social and economic benefit to the community in an environmentally responsible manner." The May 6, 2008 Ford press release title accurately describes the result, "Ford Earns Award for Turning Brownfield Green."

THE GLOBAL TRENDS

We have seen several examples of brownfield sites being developed green. Let's now examine the global trends that are pushing not only the development of brownfield sites but the greening of these sites.

A macro view of what has transpired over the past several decades in how developers, communities, and industries have moved from the "cleanup" of contaminated sites to sustainable development of contaminated properties is illustrated in Figure 1.1. Along with a move toward incorporating sustainable development practices into cleaning up contaminated properties, increased value is created in the process.

Illustrations of how value can be created by greening brownfield sites are available in the U.S. and the UK. Currently the U.S. Environmental Protection Agency (USEPA) has a brownfield initiative involving eight locations around the U.S. The initiative focuses on redeveloping the locations with green buildings.[2] In the UK the best example is the BedZED Project. This UK "carbon neutral," zero energy housing community was developed on a brownfield site, formerly a wastewater treatment plant. The site now holds 82 residential homes with private gardens and 14 apartments as well as green space.[3]

The trends that are driving the greening of brownfield sites include a mix of increased urbanization, increased awareness of the benefits of green building practices, and resource scarcity. Each of these trends is briefly examined to provide a feel for the global business drivers in play.

Urbanization and Population

Although many trends are increasing the move toward the greening of brownfield sites, the key driver is the global move of rural populations to urban settings. For the first time in history a greater percentage of the global population lives in an urban setting as opposed to rural communities.

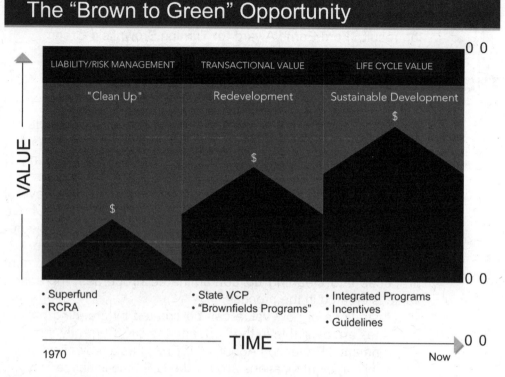

Figure 1.1   The greening brownfields opportunity.

This is truly a global movement as illustrated in Figures 1.2, 1.3, and 1.4. Reference: http://esa.un.org/unup/index.asp. (Statistics are from the Population Division of the Department of Economic and Social Affairs of the United Nations Secretariat, World Population Prospects: The 2006 Revision and World Urbanization Prospects: The 2007 Revision, http://esa.un.org/unup.)

This movement of the global population into urban settings is forcing the development of infill properties in developed countries such as the U.S. and Europe and explosive growth of cities in China and India.

Green Building

The green building movement has exploded over the past several years and is now becoming mainstream. The success of the USGBC illustrates the increasing success of the green building movement

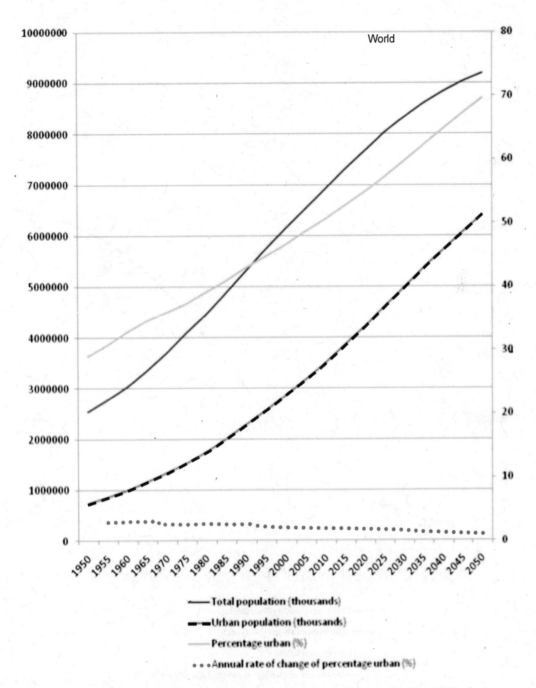

Figure 1.2 World population and percentage urban population growth since 1950. *Source: World Urbanization Prospects: The 2007 Revision Population Database United Nations.* Reprinted with permission.

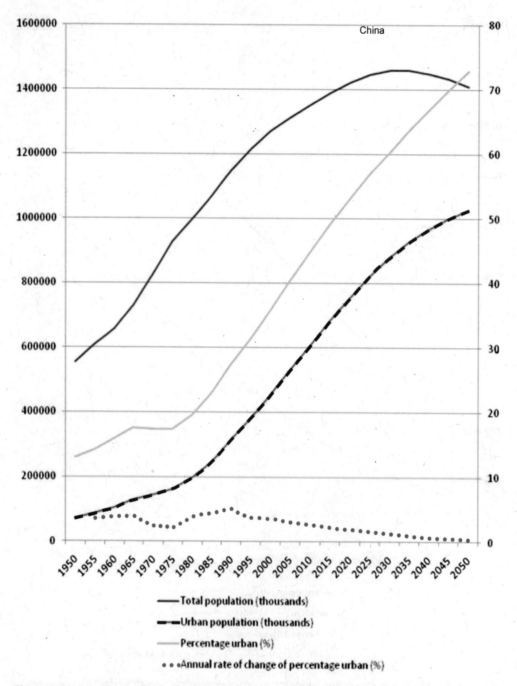

Figure 1.3 Chinese population and percentage urban population growth since 1950. *Source: World Urbanization Prospects: The 2007 Revision Population Database United Nations.* Reprinted with permission.

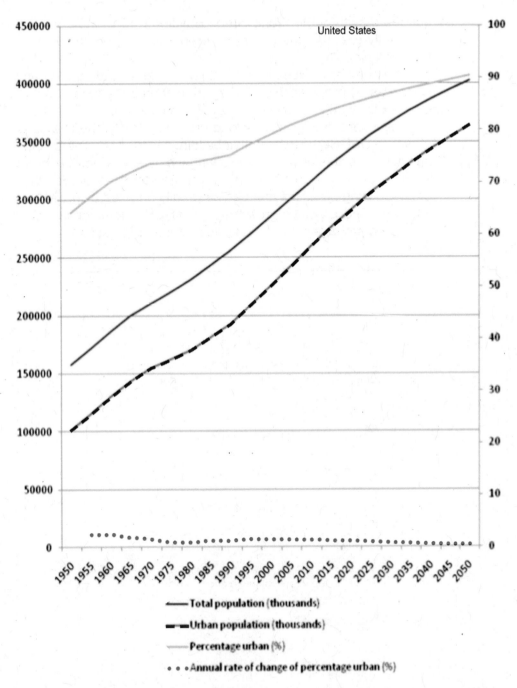

Figure 1.4   United States population and percentage urban population growth since 1950. *Source: World Urbanization Prospects: The 2007 Revision Population Database United Nations.* Reprinted with permission.

and the growth of the World Green Building Council (WGBC). We will briefly examine the growth of green building practices in the U.S. and globally.

The number of cities in the U.S. with green building programs is increasing dramatically. Figure 1.5 indicates the extent of various LEED initiatives in the U.S. as of July 1, 2009.

The increase of green building programs tracks the overall increase in USGBC LEED certified buildings in the U.S. The USGBC has been successful in communicating the overall benefits of green building practices and has created a standardized approach to green building practices that can be readily adopted by project developers. The dramatic growth of the USGBC commercial LEED registered projects in the U.S. is illustrated in Figure 1.6.

Various LEED Initiatives
including legislation, executive orders, resolutions, ordinances, policies, and initiatives are found in

43 states, including

190 localities (126 cities, 36 counties, and 28 towns),

33 state governments,

13 federal agencies or departments,

16 public school jurisdictions, and

39 institutions of higher education across the USA

Figure 1.5   Various LEED initiatives in the U.S. as of July 1, 2009. *Source: http://www.usgbc.org/DisplayPage.aspx?CMSPageID=1779.*

While the growth of the green building movement in the U.S. has been significant, it is not revolutionary or unusual. Instead, it has been part of the overall global trend toward the adoption of green building practices. Examples of global "green building councils" are provided below to illustrate the globalization of green building practices. (http://www.worldgbc.org/images/stories/pdf/WGBC%20at%20UNEP%20SBCI.pdf)

Green Building Council Australia

- 310 members in 2007 (187 in 2006)
- 26 Green Star Certified buildings in 2007 (eight in 2006)
- 138 Green Star Registered buildings in 2007 (36 in 2006)
- 2800 Green Star course participants (1300 in 2006)
- 900 Green Star accredited professionals (400 in 2006)

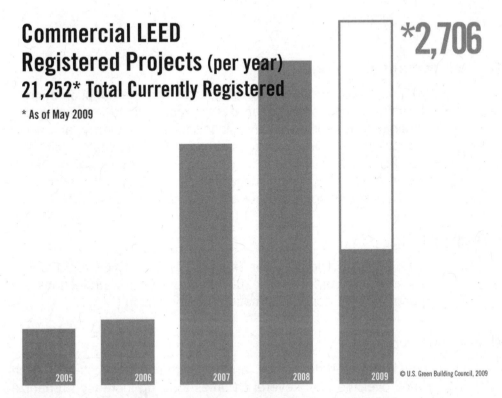

Commercial LEED Registered Projects (per year)
21,252* Total Currently Registered

* As of May 2009

*2,706

2005 2006 2007 2008 2009   © U.S. Green Building Council, 2009

Figure 1.6 Commercial LEED registered projects per year as of May 2009. *Source: http://www.usgbc.org/DisplayPage.aspx?CMSPageID=1720.*

Taiwan Green Building Council

- Started in 1999
- 1167 green building projects submitted for certification as of March 2007
- 168 certified green buildings as of March 2007

New Zealand Green Building Council

- Started in 2007
- Estimated 25 projects sought certification at that time

United Kingdom Green Building Council

- Started in 1990
- 285,250 green buildings registered for green building "compliance"
- 73,125 projects certified

Global Resource Constraints

In general, the reuse of brownfield sites and the adoption of green building practices reflect an overall response to increasing constraints on resources such as land, materials, energy, and water. No longer are these resources abundantly available to businesses and the public sector. One only has to look at increasing global demand for energy, land, and water to appreciate the strains the availability of resources have on real estate development.

Energy Demand

The increasing demand for energy (The Energy Outlook) is illustrated in Figure 1.7 in the 2008 energy demand projections prepared by the Energy Information Administration (EIA).

Developed Land

The increase in developed land within the U.S. has steadily escalated over the past several decades tracking increased urbanization. These trends are illustrated by U.S. Department of Agriculture (USDA) studies shown in Figures 1.8 and 1.9.

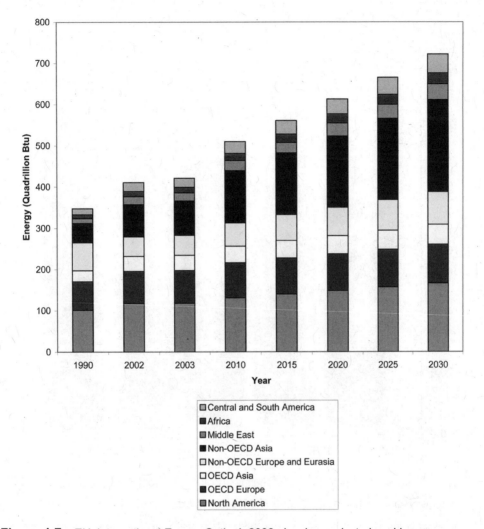

Figure 1.7 EIA International Energy Outlook 2008 showing projected world energy consumption. *Source: Energy Information Administration.* (http://www.eia.doe.gov/oiaf/aeo/pdf/trend_2.pdf)

VALUE CREATION—BROWNFIELDS REMEDIATION AND GREEN BUILDING

The movement of organizations focused on the development of brownfield sites in promoting green development can be illustrated by the U.S. National Brownfield Association's effort to influence the USGBC to increase the LEED point value for developing a brown-

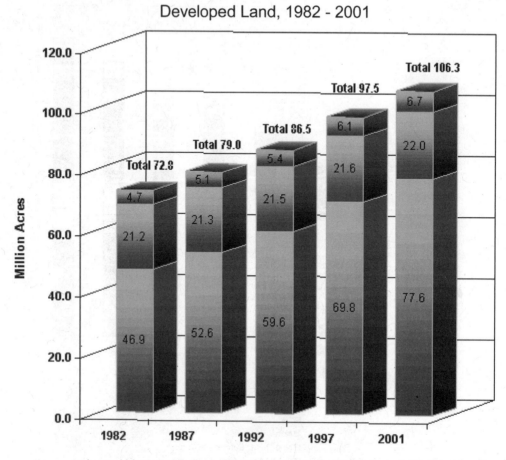

Developed Land, 1982 - 2001

Figure 1.8 2001 Annual Natural Resource Inventory by NRCS. *Information provided by the National Resources Conservation Service does not constitute endorsement by the U.S. Department of Agriculture of any commercial projects or services.* (http://www.nrcs.usda.gov/technical/NRI/2001/nri01dev.html)

field site from 1 point to 5 points (http://www.brownfieldassocia-tion.org). If this effort is successful it will encourage private and public sector enterprises interested in achieving LEED certification to build on a brownfield site as opposed to greenfield development.

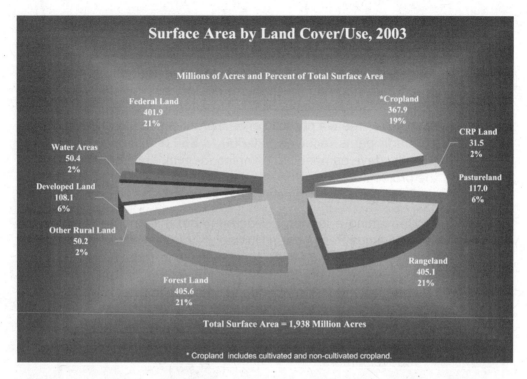

Figure 1.9 2003 Annual Natural Resource Inventory by NRCS. *Information provided by the National Resources Conservation Service does not constitute endorsement by the U.S. Department of Agriculture of any commercial projects or services.* (http://www.nrcs.usda.gov/technical/NRI/2003/nri03landuse-mrb.html)

Both the USEPA and the National Brownfield Association (NBA) are working to promote the greening of brownfield sites. In August of 2008 (Architecture & Design, August 11, 2008) the USEPA provided $500,000 to green brownfield sites. Sixteen projects were selected nationally to receive funding for technical support under a USEPA Brownfields Sustainability Pilots program. Projects included a former California lumber mill, building a solar plant on a Texas landfill, and the salvaging of recyclables from textile mills in Alabama.

According to Susan Bodine, the assistant administrator for USEPA's Office of Solid Waste and Emergency Response, "Brownfields redevelopment and sustainable reuse can go hand in hand."

I believe it is more than "going hand in hand"; it is creating enhanced value for developers and communities. Moreover, only remediating a brownfield site leaves opportunity and value on the table.

The goals of the USEPA funded pilot projects are to support the reuse and recycling of construction and demolition materials, green building and infrastructure design, energy efficiency, water conservation, renewable energy development, and native landscaping to create a portfolio of best practices. Each of the pilot project participants receives $20,000 to $50,000 to incorporate sustainable development into planning, design, and implementation of the project.

Representative projects include:

- Converting a former 171-acre lumber mill and sash factory on the Samoa Peninsula in Humboldt County, California, to a green mixed-use commercial, residential, and recreational development that also preserves the historic character of the area. The county wants to incorporate Leadership in Energy and Environmental Design (LEED) Neighborhood Development Principles in the project, and the EPA's $50,000 in assistance will help support green design and eco-friendly standards sought for the redevelopment.

- Construction of a solar plant on a former 300-acre landfill in Houston, Texas. The USEPA is providing $50,000 in assistance for analysis of the environmental, engineering, and energy issues associated with the project.

Let's take this a step further. USEPA recognizes that there is value in incorporating sustainable development practices and "green practices" into the development of brownfield sites. Couple this with a drive to build "zero-net energy" commercial buildings recently announced by the U.S. Department of Energy (U.S. DOE) (Architecture & Design, August 6, 2008) and we can get a sense for where development projects are likely headed.

The USDOE and five national laboratories are working together to accelerate the development of marketable zero-net energy commercial buildings. Think about it, remediation to zero net energy. "No contamination" and no net energy use. The value from taking a liability and turning it into not only an asset, but a development that has a dramatically reduced life cycle operating cost is truly innovative.

A National Laboratory Collaborative on Building Technologies (NLCBT) and the USDOE Zero-Net Energy Commercial Building

Initiative (CBI) were established under the U.S. Energy Independence and Security Act of 2007 (EISA) signed into law in December 2007 calling for all new commercial buildings to be so efficient in energy consumption and in on-site renewable energy generation that they offset any energy use from the grid. The target date for that goal is 2025. The initiative sets a zero-net energy goal for all commercial buildings by 2050.

The mission of the collaborative, a joint project of the USDOE's Building Technologies Program and the five national labs, is to devise a multi-year plan, timetable and strategies for achieving the 2025 goal for new commercial buildings and to jump start the transfer of new technologies from research and development to the marketplace.

REFERENCES

1. Databank, Inc.—an Atlanta commercial real estate research company.
2. Colangelo—Atlantic site 2003.
3. http://www.epa.gov/brownfields/html-doc/greenbld.htm.

Convergence—
The USEPA Meets
the USGBC

Of all the drivers in the perfect storm creating opportunities to green brownfield sites, the two most significant in the U.S. are the USEPA and the USGBC. In some ways an unlikely pair, but nevertheless both are moving toward a convergence in understanding that promoting greening brownfield sites creates increased value for the public and developers and makes good environmental sense.

If a brownfield is developed but not built green, or at the very least aspects of sustainable development are not incorporated into redevelopment, value from the site is not completely captured. In other words, the highest value was not created from redevelopment.

The U.S. currently represents the clearest example of the convergence of traditional environmental programs with sustainable development. Historically these two initiatives have lived in separate worlds. Decades of environmental programs starting with the federal Superfund program (http://www.epa.gov/superfund) coupled with state voluntary cleanup programs (VCUP) were primarily focused on the protection of human health and the environment and not necessarily on beneficial use of the property, let alone sustainable development.[1]

This has changed rapidly with both the USEPA and the USGBC recognizing the importance of redeveloping urban infill sites (most of which have some environmental impact). Although each organization approaches this differently, they both bring enormous experience and talent to addressing the need to revitalize contaminated urban sites.

We will examine how these two perspectives together have converged and are transforming the redevelopment of contaminated properties.

THE DEFINITION OF BROWNFIELDS

First let's examine how the "green building world" and the "brownfield world" define brownfields.

According to Wikipedia, brownfields are "abandoned, idled, or under-used industrial and commercial facilities available for re-use where expansion or redevelopment may be complicated by real or perceived environmental contaminations."

In practice, "brownfield land is land previously used for industrial purposes, or certain commercial uses, and that may be contaminated by low concentrations of hazardous waste or pollution, and has the potential to be reused once it is cleaned up. Land that is more severely contaminated and has high concentrations of hazardous waste or pollution, such as Superfund or hazardous waste sites, do not fall under the brownfield classification."

The USGBC definition is "a site documented as contaminated (by means of an ASTM E1903-97 Phase II Environmental Site Assessment or a local Voluntary Cleanup Program) OR a site defined as a brownfield by a local, state, or federal governmental agency."

The USEPA definition is that a brownfield site means "real property, the expansion, redevelopment, or reuse of which may be complicated by the presence or potential presence of a hazardous substance, pollutant, or contaminant."

In the USEPA's view, "the cleaning up and reinvesting in these properties takes development pressures off of undeveloped, open land, and both improves and protects the environment. Costs of these lands can be considerably cheaper, and many contractors looking to expand their own operations can find good deals in great locations if they are willing to take on the cleanup aspect. There can be federal and local incentives to remediate the toxic areas and redevelop the site. Remediation involves the removal of hazardous material from the site's groundwater and soil."

THE USEPA PERSPECTIVE

Let's start with the USEPA. Although we will get into greater detail in subsequent chapters, it is worth noting recent USEPA initiatives illustrating a change in thinking by the agency.

In July 2008, the USEPA announced plans to provide $500,000 in technical assistance to 16 brownfield sustainability pilot projects. In other words, greening brownfield sites.[2]

This is a significant expansion of USEPA's mission to remediate contaminated properties and in many ways aligns USEPA with the goals of the USGBC. These redevelopment programs by USEPA are now factoring in benefits such as the creation of "green jobs," community needs, and overall social performance.

An overview of the pilot projects are as follows.

Lynchburg, Virginia. The Allen Morrison property has been an industrial site for over 100 years, including chemical manufacturing. The USEPA selected the site based on public health concerns and recognition that additional recreational facilities were needed within the community.

The USEPA support will focus on proactive engagement of the community in the pilot project, general support for sustainable development best practices for the construction of the public park, and promoting "active living."

Cleveland, Ohio. The City of Cleveland bundled USEPA funding along with other funding sources to investigate and remediate several vacant and underutilized sites within the Cleveland area. Several of the sites require the deconstruction and recycling of building material. The focus of the USEPA grants are to evaluate public policy issues, develop best practices, and move forward with redevelopment plans for several sites.

Oklahoma City, Oklahoma. USEPA is supporting the redevelopment of a site in the Capital Hill area of Oklahoma City through a partnership with the Latino Community Development Agency (LCDA). The mission of LCDA is to improve the quality of life for the Latino population in Oklahoma City. The site selected for redevelopment is a former commercial retail property vacant since the

1980s. The pilot project is designed to transform the former retail property into a community resource. USEPA's focus will be to support the design of a community rooftop garden and a sustainable stormwater management system. The project will highlight innovative design strategies while housing LCDA's new office which supports various community services such as healthcare, daycare, senior services, and other social services.

Detroit, Michigan. In Detroit, the USEPA is working with Focus: HOPE to revitalize the Oakman East neighborhood. The neighborhood includes former industrial and commercial operations, including fabrication, warehouses, and a service station. The Oakman East neighborhood was the subject of the 2008 National Brownfields Conference design charrette. The plan for redevelopment is to implement several recommendations identified as part of the design charrette.

Funds will be used for planning, green building design, stormwater management, green infrastructure design, and native landscaping. The redevelopment plan envisions commercial, park and greenway space, and affordable housing.

Focus: HOPE is a civil and human rights nonprofit organization in Detroit, Michigan, whose programs include education, training, and economic development initiatives for the local community. The organization plans to revitalize the Oakman East neighborhood.

Portland, Oregon. USEPA is funding the Delta Sigma Theta Sorority in Portland to convert an abandoned service station into a community center and transitional housing. The Delta Sigma Theta Sorority is a nonprofit organization that provides public services to the African American community in Portland.

The redevelopment site was originally evaluated under the Brownfields Showcase Communities Initiative but requires additional evaluation for sustainability features which will be funded as part of the USEPA program. The USEPA funding will support the development of infrastructure design plans, green building, stormwater, and green space. An innovative element of the green building will be the use of surplus cargo containers for the transitional housing.

Greenville, South Carolina. The Greenville Workforce Housing Project in Greenville will be funded by the USEPA to redevelop a former service station and adjacent supermarket into an affordable workforce housing project. Remediation issues at the site include underground storage tanks, and demolition and disposal of asbestos and lead paint. USEPA funding will focus on incorporating sustainability elements such as porous surfacing, rain gardens, and green building practices. The project will incorporate green building into affordable housing.

Houston, Texas. This pilot project in Houston will revitalize a former 300-acre landfill site to include ecosystem revitalization, economic development, and restoration of a historical area. USEPA funding will focus on building a solar energy installation on the former landfill and financial feasibility studies to support the development of a local market for a solar energy system and related job creation.

Boston, Massachusetts. The Jackson Square area of Boston (intersection of Roxbury and Jamaica Plain neighborhoods), which includes soil and groundwater contamination and urban fill, will be redeveloped. Redevelopment will include green building, stormwater management, and green infrastructure. Urban Edge and its partners for the "Jackson Square Redevelopment Initiative" will receive a cost effective green roof system for early phase building construction. The green roof system is viewed as a critical part of the larger effort for conversion of the underutilized land into a mixed-use development with embedded sustainability practices.

Valley, Alabama. Two former textile mills (Langdale and Riverdale Mills) in Valley were selected for redevelopment. The two former mills operated from the 1860s to the 1990s and were the primary economic drivers in the community. The USEPA is not only funding the remediation of these two mills but the community is also receiving USEPA Smart Growth Training.

Funding will be used for sustainability planning and to create an inventory of onsite materials that can be used for reuse and recycling. They city owns several abandoned mills and will promote sustainable development by inventorying mill content, evaluating the potential for reuse and recycling, and developing best practices to assist with other mill redevelopment projects.

Laredo, Texas. The Laredo Environmental Services Department will receive a USEPA grant to reclaim a contaminated wetland area, Killam Lake, along Chacon Creek, a tributary of the Rio Grande River. The city will build a new recreation center adjacent to the brownfield remediation site which will incorporate energy and resource conservation building features, an innovative stormwater management system to reduce impervious surface areas, and the use of native vegetation to conserve water resources. This revitalized brownfield site will be an essential part of the city's master trail system along Chacon Creek. The goal is for the recreation center and its surrounding environment to be a hub of activities for the local community, including canoeing, catch and release fishing, hiking and nature trails, with the overall themes of sustainability, conservation, and healthy living.

Burlington, Vermont. USEPA funding of the Moran Center in Burlington is a public-private partnership that will incorporate the reuse of a former coal-fired electric generating plant on Burlington's waterfront area. Environmental issues include the former use of the site as a bulk petroleum storage area. The Moran building at the site will be renovated and will follow LEED practices. These green building practices will include the use of green roof technologies, alternative energy, energy efficiency, and conservation practices. USEPA funding will also promote stormwater management, green infrastructure, and wetland rehabilitation. The overall goal of the redevelopment will be to create a year-round recreational and educational area to enhance access to Lake Champlain and expand the waterfront park. The Moran development will also be a catalyst for ecological enhancements of the northern portion of the downtown waterfront area.

Samoa Peninsula, California. In Humboldt County the USEPA has funded the redevelopment of the Samoa Peninsula which is a 171-acre former lumber mill and sash factory. The site contains high concentrations of lead and other chemicals related to the operation of the former mills. The USEPA funding will go toward the conversion of the former lumber mill into new commercial, residential, and recreational uses. The county wants to incorporate LEED practices into the project to develop the area in a sustainable manner and retain the historic nature of the site while promoting new businesses to increase local employment and revenue.

San Juan County, Colorado. The Anvil Mountain Neighborhood property in San Juan County is being funded by the USEPA to remediate lead and arsenic contamination from former silver, lead, and gold processing. The county is using state and USEPA funds to implement green building design, stormwater management, green infrastructure, and incorporate native landscaping practices on former mining and ore processing sites. The funding is to support the development of preliminary plans and design reports mapping out how the city will achieve green building certification for green infrastructure and green affordable housing.

Moreover, the city will identify best practices in maximizing the use of native vegetation, the productive use of runoff, and reducing the use of fossil fuels for heating and utilities under difficult conditions.

Portland, Oregon. The Tabor Commons site in Portland will be funded by the USEPA to redevelop a former gas station into a community center and a facility to promote training for "green collar" jobs. The project will be led by Oregon Tradeswomen, a community training nonprofit with funding from the USEPA, the City of Portland, and the State of Oregon. Funding will support the remediation of the site along with incorporating green building practices and the development of training materials for the "green jobs" program. The property is owned by SouthEast Uplift, a community-based organization that will also operate the community center. Oregon Tradeswomen has also received a Brownfields Job Training grant and will use the project to develop best practices that can be used for other redevelopment projects.

Allentown, Pennsylvania. The Waterfront is being funded by the USEPA and developed by the Dunn Twiggar Company, LLC. The Waterfront is a 26-acre brownfield site along the Lehigh River in Allentown. The site was a former iron and steel manufacturing site containing contaminated soil and groundwater. The USEPA funding will be used for reuse and recycling of construction and demolition materials as well as low impact practices for stormwater management and surface water protection (Lehigh River). The goal of the project is to transform the former brownfield site into a mixed-use residential, commercial, and retail development incorporating sustainable development best practices. The site also includes public space along the riverfront and access to historic and cultural aspects of the Lehigh River. An essential aspect of the redevelopment is to transform a former industrial area into a viable economic development for the surrounding community.

This new vision for the USEPA was well captured by Susan Bodine, assistant administrator of the USEPA Office of Solid Waste and Emergency Response, as "brownfields redevelopment and sustainable reuse go hand in hand" and "these pilot projects will demonstrate best practices that can be used by other communities."

These USEPA pilot projects will individually receive between $20,000 and $50,000 in assistance with the requirement that the developers work closely with communities to embed sustainable development elements into the projects.[3]

The overall funding for these projects was authorized as part of the January 2002 Small Business Liability Relief and Brownfields Revitalization Act. Objectives of the act were to increase funding and provide expanded authority and added liability protection for communities developing brownfield sites.

THE USGBC PERSPECTIVE

So where does the USGBC stand on the issue of promoting the development of brownfield sites? Under the current USGBC LEED program, one point is awarded for a green building on a brownfield site.[4] However, the LEED scorecard is being revised and may include a 110 point scorecard with more emphasis for developing on a brownfield site.[5]

The LEED scorecard version 2.2 awards one point for developing a brownfield site and a total of 14 possible points for a sustainable site out of 69 possible points, or 21 percent of the possible points. While all of the sustainable site points are not directly linked to brownfield redevelopment, they involve elements that are consistent with brownfield sites, development density, access to public transportation, and site selection.

The updated LEED 2009 scorecard keeps the credit for redeveloping a brownfield site the same, but expands the scoring of related aspects such as access to public transportation and development density. The new scorecard increases the available sustainable site points from 14 to 26, raising the value of a sustainable site to 24 percent of possible points. In fact, sustainable site points now represent 65 percent of the points required to reach LEED certified status for a new building, while in the past they only represented 54 percent. This demonstrates the rising importance of site selection as seen by the USGBC.

THE GLOBAL VIEW

It is not just in the U.S. where we see this convergence between traditional brownfield remediation and green building and sustainable development trends. An excellent example of this convergence within a global setting is the World Business Council for Sustainable Development (WBCSD) case study of the Swire Properties and Hongkong Land development project in Hong Kong.[6]

Prominent developers Swire Properties and Hongkong Land teamed in 1995 and committed to building green in Hong Kong. There was a recognition that what was really needed was to not merely build according to international best practice, but to create incentives for the property industry in Hong Kong to build green.

The result was the creation of the Hong Kong Building Environmental Assessment Method (BEAM). The WBCSD's regional partner and the developers developed the BEAM Society in 2000 along with developing guidelines. It is worth taking a closer look at the WBCSD regional partner in Hong Kong, the Business Environmental Council (BEC), as it is a useful model for public/private partnerships. The BEC was started in 1989 by 16 companies to form the Private Sector Committee on the Environment (PSCE). The PSCE recognized that the quality of Hong Kong's environment was "critical" to maintaining a "good" business environment. The PSCE led to the creation in 1991 of the Centre for Environmental Technology (CET) to drive collaboration. In 2000 the PSCE and CET merged to form the BEC and in 2003 became a regional partner of the WBCSD.

The challenge was to influence Hong Kong's estimated 70 million square metres of property space. In addition, nearly 2 billion square metres of building space is added annually.

The BEAM initiative provides a systematic approach to embedding green building practices into the planning, design, construction, operation, management, and marketing of buildings. In summary, the benefits of BEAM are as follows.

- Cost savings from more efficient use of energy and resources, in both construction and use of buildings
- Better buildings, which provide healthy and productive accommodation
- Reduced risk through assurance that best practice management is achieved
- Effective markets, as companies are able to give assurance of the green credentials of their buildings, and tenants and buyers are able to communicate their preferences
- Regulatory preparedness for both local and international standards. Hong Kong is somewhat unique in that while globally buildings account for approximately 30 percent of greenhouse gas emissions, Hong Kong is closer to approximately 70 percent (air-conditioning, lighting, and ventilation).

It was two members of PSCE, Hongkong Land and Swire Properties, who came together to address real and perceived barriers to building green in Hong Kong. Both of these developers already were committed to and had a track record for green design and construction. These developers were impressed with the UK BRE Environmental Assessment Method (BREEAM).

According to Swire Properties' Stephen Fong, they were driven by direct cost savings, corporate citizenship, and future-proofing considerations: "Being a good corporate citizen, we believe that industry should take the lead instead of waiting for legislation to tackle the issue."

In summary, BEAM defines over 100 best practice criteria to address green building for commercial, residential, and institutional buildings, as well as hotels and mixed-used complexes. As with the USGBC, separate standards are in place to appraise the environmental performance of new developments and existing premises. Briefly, BEAM focuses on:

- Site aspects (location, planning, and emissions);
- Material aspects (selection, usage, and waste management);
- Energy use (system designs and management);
- Water consumption (quality and conservation);
- Indoor environmental quality (thermal comfort, indoor air quality, lighting, noise, and vibrations); and
- Innovations (innovative techniques and performance enhancements).

The result of Hong Kong's BEAM program is that since 1996 almost 150 major property developments in Hong Kong have achieved recognition for improved performance covering over 6.8 million m^2 of space, including commercial and institutional buildings and some 36,000 residential units. BEAM's adoption since 2002 has been equivalent to approximately 25 percent of private commercial space and 10 percent of private dwellings completed each year.

SUMMARY

The convergence in thinking by public and private sector enterprises is creating previously unrecognized value from the remediation of brownfield sites. Globally, real estate developers and cities are providing the catalyst for more sustainable development projects with increased monetary and societal value.

As LEED acceptance continues to increase in the U.S., there is a need to ensure alignment of brownfield remediation and green building. The expansion, redevelopment, or reuse of brownfields may be complicated by the presence or potential presence of environmental contamination. Although reusing such property is an automatic one point toward obtaining LEED project certification, brownfield projects can inadvertently limit future green building and sustainable design with traditional application of land use controls (LUCs). The potential result is a collision of good intentions.

LUCs are a reasonable and cost-effective part of the cleanup process. Brownfield LUCs do not clean up contamination, but can responsibly manage the risk of contaminants left in place. LUCs can be engineered structures such as impervious parking lots used as caps or institutional controls such as deed restrictions prohibiting excavation.

While brownfield LUCs defer costs of a physical cleanup to preserve available capital for initial reconstruction, they may directly oppose vital construction concepts of green architecture. For instance, the use of pavements as caps over contamination to "shed" water from the property contradicts the use of permeable pavements to maximize on-site storm water retention. Parking lots can conflict with green space, water features, and ecological habitat restoration.

However, such collisions are avoidable. If we understand how and when brownfield redevelopment affects LEED, we can progress from what have been sequential, or at best parallel, pathways to a process that integrates brownfield remediation and LEED early in site planning. Sustainable design principles can be incorporated earlier in the project by making it part of the environmental assessment and early cleanup planning.

An understanding of the relationship between brownfields cleanup and green design in LEED certified redevelopment also provides opportunities for valuable LEED credits using brownfield assessment information to support green building requirements related to sustainable site planning and water resources.

"Green" Brownfields Reuse

When collisions occur, it results from a traditional misconception that green building on brownfield properties is a sequential and linear process. The reality is "an ounce of prevention is worth a pound of cure" or, in our scenario, about a half a million dollars.

Collision is avoidable. Consider a less traditional approach in which brownfield cleanup and green building are not sequential, but simultaneous—an approach where green design will enter much earlier in the property restoration process than is the norm and merge gently with the brownfield process.

Design of LUCs can consider LEED criteria. Cleanup can consider effects on construction and physical infrastructure necessary for certification.

Sustainable Redevelopment Concepts at the Assessment Stage

Brownfield projects can prepare for LEED and green design even before the two technical disciplines merge for an actual project. Environmental assessment data can be used to map a property beyond the chemistry, identifying areas that appear to best support sustainable reconstruction features. The result is a brownfield site positioned for both "green" developers, Smart Growth-related cleanup, and elevated scoring in a future LEED project.

Assessment mapping considers parameters for certification of a LEED Neighborhood Development (Pilot), specifically how site soil and contaminant conditions may affect or support Green Construction & Technology—Credit 8: Contaminant Reduction in Brownfields Remediation, and Credit 9: Stormwater Management. This provides architects/designers with three-dimensional mapping of residual contaminants for optimal placement of rain gardens or permeable pavements in areas which will not promote secondary mobility of subsurface residual contaminants to groundwater.

In areas where cost or avoidance of elevated residual contamination might have to occur during cleanup and reconstruction, assessment sustainability mapping can assist designers in identifying areas where green design can be used as part of the remedy. This might occur by designing impermeable, rainwater-harvesting structures that act as subsurface "caps" for deeper contaminated materials but allow near-surface water movement and infiltration for collection. This allows constructing a parking lot of permeable pavement that allows stormwater infiltration and reduces surface water contamination.

Essential in brownfield development and LEED convergence is integrating the planning and design phases early so that LUCs for cost-effective cleanup or contaminant management become part of green design. Avoid the collision. It is time to re-examine the current linear pathways of thinking—your brownfield development's long-term success and profitability in the new "green society" may depend on it.

REFERENCES

1. Texas: http://www.tceq.state.tx.us/remediation/vcp/vcp.html; Washington:http://www.ecy.wa.gov/programs/tcp/vcp/Vcp-main.htm; New Jersey: http://liberty.state.nj.us/dep/srp/volclean/; New York: http://www.dec.ny.gov/chemical/8442.html; Oregon: http://www.deq.state.or.us/lq/cu/voluntarycu.htm; Maryland: http://www.mde.state.md.us/Programs/LandPrograms/ERRP_Brownfields/vcp_info/index.asp;Oklahoma: http://www.deq.state.ok.us/LPDnew/VCPIndex.htm
2. July 28, 2008, http://yosemite.epa.gov/opa/admpress.nsf/dc57b08b5acd42bc852573c90044a9c4/5bb8488345b8985a85257495005dec2d!OpenDocument
3. http://epa.gov/brownfields/sustain_plts/factsheets/com_st_spfs.pdf
4. http://www.usgbc.org/DisplayPage.aspx?CMSPageID=220
5. http://www.usgbc.org/DisplayPage.aspx?CMSPageID=1849
6. http://www.wbcsd.org/Plugins/DocSearch/details.asp?DocTypeId=24&ObjectId=MjkyNzY&URLBack=%2Ftemplates%2FTemplateWBCSD5%2Flayout%2Easp%3Ftype%3Dp%26MenuId%3DMTU3OA%26doOpen%3D1%26ClickMenu%3DLeftMenu

Regulatory Drivers

In the U.S. the major drivers for the transformation of brownfield sites are both state and federal regulations. In fact, U.S. regulations have led the global movement to remediate and reuse brownfield sites. Global approaches to brownfield sites vary but all are aligned on providing the regulatory framework to effectively remediate these contaminated properties.

We will examine the regulatory framework for the remediation of brownfield sites in the U.S. (both state and federal programs) and then provide the global view. These regulatory drivers form the basis for the growth of brownfield redevelopment and the foundation for incentives to expand redevelopment opportunities.

A key point with regulations is that public policy and private enterprise should align to maximize value creation. This is not unique to transforming brownfield sites into green developments but, after decades, the brown to green movement has led the way.

U.S. REGULATIONS

Let's start with the U.S. regulatory framework. The real push for brownfield redevelopment and the program that moved industry and developers to examine how to transform liabilities into assets came from the USEPA in the form of the Comprehensive Environmental Response, Compensation, and Liability Act, 1980 (CERCLA or Superfund) and the Resource Conservation Recovery Act (RCRA). Moreover, it provides the USEPA with the authority to identify "potentially responsible parties" and obtain their cooperation (funding) in the remediation through consent orders or consent decrees.

The USEPA can remediate "orphan sites" where potentially responsible parties (PRPs) can not be identified. They can also recover costs from financially viable individuals and companies after a site has been remediated, and they are authorized to implement CERCLA (http://www.epa.gov/lawsregs/laws/index.html) in all 50 states and U.S. territories.

The USEPA also has authority under RCRA (1976 and the 1986 amendments to RCRA) to "address" environmental contamination resulting from underground tanks storing petroleum and other hazardous substances. In addition, the Hazardous and Solid Waste Amendments (HSWA) of the 1984 amendments to RCRA addressed corrective action for releases (http://www.epa.gov/lawsregs/laws/index.html).

Based upon the experience of the early days of the program, and there were numerous lessons to be learned, CERCLA was amended in 1986 through the Superfund Amendments and Reauthorization Act (SARA) which reauthorized the USEPA to continue the program and included amendments, clarifications, technical requirements and additional enforcement authorities (http://www.epa.gov/lawsregs/laws/index.html).

Further improvement (some would argue that CERCLA has a long way to go to be truly effective) with CERCLA came with the Small Business Liability Relief and Brownfields Revitalization Act of 2001 which provided funds to remediate brownfield sites and clarified liability provisions for innocent landowners (http://www.epa.gov/swerosps/bf/sblrbra.htm).

While CERCLA and RCRA are the backbone of U.S. federal regulations governing the remediation of contaminated property, several other federal regulations are significant. These are briefly summarized below:

- USEPA—Oil Pollution Act (OPA) of 1990 provides for the prevention and cleanup of oil spills which have the potential to impact the waters of the U.S. (http://www.epa.gov/lawsregs/laws/index.html).

- USEPA—Pollution Prevention Act (PPA) of 1990 includes practices that increase efficiency in the use of energy, water, or other natural resources, and protect the resource base through conservation (http://www.epa.gov/lawsregs/laws/index.html). With the advent of sustainability this federal regulation was, in my opinion, ahead of its time and addresses proactive management of natural resources.

- Other federal regulations—Although not strictly regulatory in nature, several programs were established to provide research on air and water quality and to assist in establishing national standards. Although not perfect, these programs moved forward in establishing much needed federal standards.

  These programs include: the Federal Water Quality Administration (FWQA), formed in 1965; The National Air Pollution Control Administration (NAPCA), originated as a research entity in 1955 and renamed as the NAPCA in 1968, initiating air and water quality standards by the mid-1960s.

  Both FWQA and NAPCA were at first part of the Public Health Service. The FWQA broke off from the PHS in 1966 and became part of the Department of the Interior. The USEPA was formed in 1970 and became the vehicle for an approach to addressing contamination in an integrated manner.

  The creation of the USEPA came on the heels of the publishing of Silent Spring by Rachel Carson in 1962. The challenges for USEPA at the time were to balance ecological goals, the interpretation and communication of technical environmental information to the public and lawmakers, and the uncertainty associated with environmental science at the time.

  The challenges confronted by the USEPA included addressing DDT, multi-media pollution, and abandoned hazardous waste sites. The USEPA has made significant progress in addressing contaminated properties through integrated approaches to brownfield remediation and redevelopment. (http://www.epa.gov/history/topics/regulate/01.htm).

U.S. STATE REGULATIONS

Environmental regulations at the state level, in general, closely track federal regulations. In particular, this is the case for state equivalents of CERCLA and RCRA focused on the management, storage, transportation, and remediation of hazardous substances along with liability provisions.

Following are some general comments regarding state programs:

- States typically provide "flexible" regulatory programs to encourage voluntary cleanup (http://www.epa.gov/brownfields/ partners/finan_brownfields_epa_print.pdf).

- State and tribal programs play a significant role in cleaning up brownfield sites. Congress had the view that "The vast majority of contaminated sites across the nation will not be cleaned up by the Superfund program. Instead, most sites will be cleaned up under state authority." Moreover, there is a framework for negotiations between USEPA and states regarding voluntary cleanup programs (VCUP). This initial framework was the basis for current programs guiding cooperation between the USEPA regional offices and tribe and state agencies. The USEPA also provides funding for tribal and state response programs as grants.

  These USEPA funded state and tribal programs focus on the investigation, remediation, and redevelopment of brownfield sites. These funds are specifically designed to support programs to enhance current programs and capabilities, and in some instances develop new programs. Funding has also been used to establish a revolving fund for remediation, fund oversight activities, purchase insurance as needed, or the funding of other project financing mechanisms. (http://www.epa.gov/ swerosps/or/vcp_hist.htm).

- Risk-based cleanup standards are in place in almost all state programs which consider the end use of the property (residential, commercial, or industrial) and any land use restrictions (http://www.epa.gov/swerosps/bf/pubs/st_res_prog_report.htm). Moreover, these programs have moved from mere remediation programs to remediation coupled with resuse of the contaminated properties.

- Several states have innovative programs, with Virginia, California, and Pennsylvania as excellent examples. http://www.ncsl. org/programs/environ/brownfields/brwnfieldspresent0904.htm. Pennsylvania has a "Land Recycling program," California has the "Site Mitigation and Brownfields Reuse Program" and Virginia has the "Brownfields Land Renewal" program. These state programs have been successful in promoting the redevelopment of brownfield sites and economic development.

Now for a view of brownfield programs in Europe.

THE EUROPEAN UNION

Overview

The European Union (EU) has undergone a significant transformation over the past several decades in addressing brownfield programs. In some respects there has been a "role reversal" between the U.S. and the EU concerning environmental leadership (Vogel and Kelemen, MS; Sbragia and Damro 1999).

During the early 1970s the U.S. was a leader in environmental issues, with this leadership continuing through the early 1980s. A shift in leadership occurred during the late 1980s and early 1990s when environmentalists gained political influence in European countries (Mair 2001).

What has happened is that the EU has developed comprehensive environmental regulations ranging from rules on air and water pollution, to waste management and recycling, to genetically modified organisms (GMO), to chemical safety regulation. (http://www.princeton.edu/~smeunier/Kelemen.doc).

As part of this shift the EU enacted regulations to address brownfields remediation and redevelopment. In the EU the estimated number of contaminated sites is 3.5 million (http://ec.europa.eu/environment/soil/pdf/com_2006_0231_en.pdf). A brief summary of these regulations is provided below.

- Directive 2004/35/EC of the European Parliament and of the Council of 21 April 2004 on environmental liability with regard to the prevention and remedying of environmental damage. The objectives of this first European Communities (EC) legislation included the application of the "polluter pays" principle and the establishment of a common framework for liability with a view to preventing and remedying damage to animals, plants, natural habitats, and water resources, and damage affecting the land. The directive's liability scheme applies to specific occupational activities and to other activities in cases where the operator is at fault or negligent. Authorities are also responsible for ensuring that the operators responsible take or finance the necessary preventive or remedial measures themselves (http://europa.eu/scadplus/leg/en/lvb/l28120.htm).

- Proposed Framework Directive for a Thematic Strategy for Soil Protection under the Commission of the European Communities (COM[2006]231, dated 9-22-06). This EU directive provides a means for ensuring a comprehensive approach to soil "protection." EU member states are required to take specific actions but the directive leaves some freedom on how to implement the requirement. The directive also provides for a common definition of contamination, its application by member states, and a list of potentially polluting activities. Member states are required to identify the contaminated sites in their territory and establish a national remediation strategy based on sound and transparent prioritization of the sites to be remediated, aiming at reducing soil contamination and the risk caused by it, and including a mechanism to fund the remediation of "orphan sites." There is a provision for the obligation for the seller or prospective buyer to provide the administration and other party in the transaction a soil status report for sites where potentially contaminating activity has taken or is taking place (http://ec.europa.eu/environment/soil/pdf/com_2006_0231_en.pdf).
- Regulatory Environment in the EU related to contaminated sites. The EU has issued several directives as outlined below (http://www.iccl.ch/download/meeting_stockholm/Session%20G%20Joop%20Vegter.pdf):
  - Ground Water Directive
  - Soil Framework Directive
  - Waste Framework Directive
  - Strategy on Sustainable Use of Resources
  - Strategy on Waste Prevention and Recycling
  - Structural Funds
  - Strategy on Soil Protection
  - Strategy on Urban Environments
  - Landfill Directive
  - Liability Directive
  - Guidelines for State-aid
  - Water Framework Directive

EU and individual country policies are now promoting the regeneration of derelict and underused sites. While there are no specific brownfield regulations anywhere in the EU, members regulate through routine planning programs. Moreover, individual countries have developed approaches to ensure the "regeneration" of urban areas which include large tracts of brownfield sites (http://www.eugris.info/Policy.asp?e=93&Ca=1&s=None&Cy=9&Co=3&Gy=111&Title=Brownfields&en=z).

EU Country Regulations

The major countries within the European Union (including Austria, Belgium, Czech Republic, Great Britain, France, Germany, Hungary, Italy, The Netherlands, Poland, Spain, and Sweden) all have some form of statutes with regard to environmental regulations and remediation of contaminated properties. Although there is significant variability in how each country implements and enforces its laws and assigns liability, there are some generalizations that can be made. Cleanup standards are usually risk based and take into account the type of land use and presence of sensitive environments. Cleanup liability can be assigned in several different manners, but it is typical for the polluter to be the primary responsible party followed by the land owner. There are some exclusions to liability which can result in government-sponsored cleanups. However, for the most part, investigation of contamination is not required prior to transfer of a property to a new owner, and transfer of liability, either through property transfer or contractual agreement is variable. (http://www.bakernet.com/NR/rdonlyres/64E6C2B2-C2FF-4B11-AA79-76E9A4DD603B/0/global_cdcontaminatedland_jun08.PDF).

ASIA/PACIFIC REGION

In the following sections I provide an overview of brownfield environmental regulations in the Asia/Pacific Region. I will spend more time examining Japan's regulations and industry as it is somewhat unique in that it is highly industrialized and is an example of a different approach to brownfield regulations than that taken by the U.S. and the EU.

Also, there are interesting developments in the People's Republic of China (PRC) based upon the nature of the government coupled with rapid industrialization and urbanization. Change will happen quickly in China in response to these trends and emerging regulations should be tracked closely.

Brownfield programs in the Asia Pacific Region vary greatly between countries with China starting to address issues such as the contamination of land and integrating sustainable land use and building practices into urban development initiatives.

An excellent overview (as of 2008) of international regulations regarding contaminated properties can be found at http://www.bakernet.com/NR/rdonlyres/64E6C2B2-C2FF-4B11-AA79-76E9A4DD603B/0/global_cdcontaminatedland_jun08.PDF.

China

In China there apparently are no specific statutes related to environmental regulation and remediation of contaminated land, and the laws that are present only vaguely address this issue. Policies regarding cleanup standards, assignment of responsibility and liability, and enforcement of cleanup actions are also vague and somewhat poorly defined.

As we have seen recently, environmental issues do get attention if there is an immediate and serious threat to human life or property, but there does not appear to be a consistent application of environmental regulation at the national, provincial, or local level.

However, there are laws in place to address soil contamination. These laws typically relate to agricultural lands (Environmental Protection Law, Land Administration Law, Environmental Quality Standards for Soil, and Environmental Quality Risk Assessment Criteria for Soil at Manufacturing Facilities). Initial progress in addressing brownfield sites was the "State Council's Decision on Implementing the Scientific Concept of Development and Strengthening Environmental Protection" (December 2005) which focuses on risk assessment and "restoration" of former industrial sites.

China's approach to the remediation of contaminated properties is, as one would expect, closely tied to its form of government. Former industrial and contaminated properties were state owned. This makes liability issues and resolving them problematic.

If the contaminated property was formerly state owned then who pays for cleanup? The state?

An interesting example of redevelopment is the City of Dalian where there is a movement to shift industrial operations out of the urban center. Industrial facilities are moving into the outlying areas leaving the inner city with residential development opportunities. In this scenario, the municipality takes the lead on ensuring that surface structures are demolished and the infrastructure is in place for redevelopment.

Bidders interested in redevelopment of the contaminated site are required to submit plans for redevelopment including an Environmental Impact Statement (EIS). The approach is rather unique in that only visible contamination is addressed in air, water, and solid waste, while non-visible contamination (dissolved volatile organic compounds, for example) is not required to be addressed.

What is remarkable about the approach is that since the developer is not liable for any contamination and because enforcement regulations are "lax," developers are typically unwilling to develop brownfield sites.

There are apparently many bidders for these sites and municipalities and the state does not need to provide incentives for remediation. Overall, the challenges are:

- Absence of historical information regarding prior operations;
- No involvement of the public as the properties are owned by the state; and
- Soil and groundwater contamination remain in place which is a potential problem since many of these sites are developed as residential sites.

We can expect all of this to evolve and for the legal, public policy, and technical issues to be resolved over time. (http://uwspace. uwaterloo.ca/bitstream/10012/3662/1/Xiaoling_Thesis_final2.pdf)

As of 2008, the PRC is drafting regulations to guide the remediation of soil and groundwater such as the "Soil Contamination Prevention and Control Law" and additional standards that apply to soil and groundwater. Other general laws include the "Environmental Protection Law," "Solid Waste Pollution Prevention and Control Law," and "Regulation on Discarded Hazardous Chemicals Law."

The PRC is also working on drafting a definition of "contaminated land" and the concept of contaminated lands only exists in approximate terms.

Despite the evolving nature of definitions and regulations in China there are some laws that address the remediation of contaminated land such as the Environmental Protection Law, Solid Waste Pollution Prevention and Control Law, and the Regulation on Discarded Hazardous Chemical Law.

With respect to cleanup standards, remediation is required when a site represents an immediate or serious threat to human life or property. A "recommended" technical standard exists which is the Environmental Quality Risk Assessment Criteria for Soil at Manufacturing Facilities HJ/T 25-1999 ("HJ/T 25-1999") and can be used as a reference for assessing the risk of the soil pollution hazard at industrial areas. This standard is recommended but not mandated.

It is my understanding that pilot projects are being conducted using risk-based cleanup standards that will consider future land use restrictions and the potential of off site migration of contaminants. This is similar to U.S. environmental regulations.

As previously mentioned there is a regulation related to soils for agricultural use. The "Environmental Quality Standard for Soils (GB 15618-1995 1996-03-01)" specifies limits of "polluting substances" such as cadmium, mercury, arsenic, copper, lead, chromium, zinc, nickel, DDT). There are no current cleanup standards established with pilot projects using risk based cleanup concentrations.

While there are no substantive cleanup standards for soils there are clean up standards for water under Article 76 of the PRC's Water Pollution Prevention and Control Law. Also, under the PRC Water Law, Article 31 and the PRC Marine Environment Protection Law, Article12, persons causing water pollution in excess of approved discharge limits are responsible for cleanup measures.

It is important to note that potential purchasers of land are advised to conduct environmental due diligence prior to purchase.

Finally and most importantly, developers need to be positioned and prepared for major changes in PRC environmental laws as they relate to property contamination. Even China appears to be moving toward more of a "strict, joint, and several liability" framework.

Japan

The approach Japan has taken with regards to brownfields regulations is to focus on addressing environmental risks from contaminated land, rather than on seeking beneficial reuse of brownfield sites. (N. Otsuka & H. Abe, http://library.witpress.com/pages/PaperInfo.asp?PaperID=18617) This has shaped the thinking of Japan public policy toward the reuse of brownfield sites (http://www.iccl.ch/download/meeting_stockholm/Session%20A%20Hiroaki%20Sato.pdf).

While Japan has addressed environmental contamination issues of the 1960s, it has only recently addressed the transactional aspects of brownfield sites.

In 1967, environmental regulations were established in response to economic growth followed by additional regulations to address air and water pollution (Miki Mitsunari). In 2002 the Soil and Contamination Countermeasures Law was enacted to address environmental assessments and remediation. This law was a trigger for Japan to develop a robust environmental due diligence market with an eye toward sustainable development.

As Japan emerged as a major economic power there was a refocus on production and relocation of manufacturing plants. This relocation resulted in an increased awareness of environmental issues.

This shift coupled with the emergence of U.S. and European firms conducting pre-transaction environmental due diligence resulted in an increased awareness by the public and private sectors. As a result, the Soil Contamination Countermeasures Law was passed by Japan's legislature and implemented on February 15, 2003.

In addition to this law, Japan amended the "Building Lots and Buildings Transaction Business Law" which requires property brokers to disclose if a property is identified by the government as a contaminated site. Furthermore, the Real Estate Appraisal Regulation was amended in 2003 to consider the presence of soil contamination on a property when determining its financial value.

Overview of Regulations

In Japan, there are statutes with regard to environmental regulations and remediation of contaminated land. Cleanup standards tend to be risk based and are enforced by the Ministry of Environment. Cleanup liability can be assigned either to the polluter or the landowner and liability can be transferred to another party through contractual agreement (http://www.bakernet.com/NR/rdonlyres/

64E6C2B2-C2FF-4B11-AA79-76E9A4DD603B/0/global_cdconta-minatedland_jun08.PDF). A brief overview of brownfield regulations in Japan follows.

The remediation of contaminated land is regulated by the Soil Contamination Prevention Law (dojo osen taisaku ho) Law No. 53 of 2002 (hereinafter "SCPL") and the Farm Land Soil Contamination Law. Contaminated land is subject to these regulations if any of 26 "specified harmful substances" are found in concentrations greater than the standards set out in the SCPL.

With regards to liability, the landowner is primarily responsible for remediation. However, the "polluter" would be liable to cleanup the site if its identity is known, if the landowner can make the case that the polluter is responsible, and the local government (prefectural governor) considers it appropriate (also considering the ability to pay).

The SCPL establishes specific standards for the 26 substances identified in the law. As with most regulations, if the concentrations exceed the standard then a remediation order would be issued. The goal of the remediation would be to protect human health, which is similar to CERCLA in the U.S. where the goal is "protection of human health and the environment."

In addition to the SCPL, local prefectural environmental ordinances may apply to specific substances as well as possibly requiring stricter cleanup standards.

While each site and remediation approach is unique, remediation to preclude direct ingestion consists of fairly standard techniques such as: preventing access, cover, in situ containment, soil excavation, in situ treatment, etc. A similar spectrum of alternatives for the treatment of groundwater is also available for consideration.

Growth of the Japanese Environmental Remediation Market

Based upon increased development the Japanese environmental remediation market is growing as is an evolution of accounting rules and investor scrutiny with regards to land development. The environmental soil remediation market has grown approximately 30 percent per year since an environmental remediation law was adopted in 2003 (Greg Rogers and Miki Mitsunari).

Japanese government statistics indicate that approximately 35 percent of former manufacturing sites in Japan are impacted by soil contamination. Due to the prevalence of mixed land use in urban areas, contamination from historical industrial activities is being discovered in commercial and residential areas with increasing frequency. Moreover, the Japanese Ministry of Environment (MoE) states that as of 2003 the value of land impacted by soil contamination is approximately $360 to $780 billion and the estimated costs of soil remediation are about $140 billion.

While currently most properties are not subject to site assessments, it is anticipated that changes in the regulatory framework would increase the number of site assessments over time. The MoE has been considering the expansion of site assessment requirements under current laws and it is expected that new regulations will be forthcoming. These new regulations will provide for more flexible risk-based remediation standards.

It is worthwhile to note that under current Japanese accounting practices the recognition of environmental liabilities and future remediation obligations is not required, although this non-recognition is apparently changing and regulations are moving towards recognition of these liabilities and potential costs as a result of progress made by the Accounting Standard Board of Japan (ASBJ). This movement in Japan is similar to FIN 47 in the U.S. and efforts in the EU to revise its accounting standards. As foreign investment in Japan increases one can expect greater acceptance to global accounting, reporting and due diligence practices. Publicly traded Japanese companies are moving toward a recognition of this convergence in environmental practices. This will, in turn, result in increased growth of the Japanese environmental remediation industry. As we have seen in the U.S., the growth of environmental industry promotes greater "comfort" in the acquisition and reuse of brownfield sites (http://www.mizuho-ir.co.jp/english/knowledge/ef0710.html) (Rogers and Mitsunari, Environmental Finance, October 2007). Current regulations in place have sparked growth in the Japanese remediation and continued growth is expected as new regulations come online. Site assessments are performed today when a "large number" of people will use a property which is mostly industrial, commercial, and multifamily buildings. However, this approach is not a standardized approach such as the ASTM E 1527 for U.S. phase 1 assessments.

As of 2004 the Japanese environmental assessment and remediation market was estimated at $1 billion. This is considered a fraction of the potential market size of $120 billion as estimated by the Geo-Environmental Protection Center (a nonprofit under MoE).

What would be considered as phase 1 and phase 2 environmental assessments conducted in the U.S. represent about $200 million of this market size. The market grew by about 50 percent annually in 2003 and 2004, while the number of phase 1 and 2 assessments expanded to approximately 6300 with costs of $130 million (U.S.). The increase in the number of assessments and corresponding market size has had a corresponding downward pressure on the costs of a phase 1 assessment as one would expect. In the U.S. phase 1 assessments are considered a commodity. Low cost with disproportionately high potential liability best characterizes these phase 1 assessments.

While there are similarities between the U.S. and Japanese markets, the differences are significant. The major difference is that Japan does not have a corresponding policy to the U.S. federal and state brownfield programs and related redevelopment incentives. In Japan the focus is on compliance and in the U.S. the focus is on redevelopment (bringing properties back to productive use ("asset conversion"). This is not to say that in the U.S. compliance is not an issue, but that the U.S. market has moved beyond compliance. Moreover, while in Japan the phase 1 and 2 nomenclature is used there is really not standardization in the processes and each assessment is somewhat unique. To say the least, this is a challenge if you are trying to understand the results of an assessment or group of assessments. Emerging issues in the Japanese market will likely include: increased standardization of the phase 1 and 2 processes; a risk-based approach to remediation; a real brownfields program to promote redevelopment; and increased local and national incentives to promote site remediation and reuse.

So, the Japanese market will mature into a market more like the U.S. and in all likelihood the market in China will move toward a Japanese and U.S. market. As the goal is ultimately to remediate and reuse sites, one can expect rapid evolution in technologies and legal frameworks. (http://www.mizuho-ir.co.jp/english/knowledge/esa0608.html)

Australia

There are an estimated 80,000 to 100,000 contaminated properties in Australia according to Jon Doumbos, managing director of Sydney-based remediation technology developer Dolomatrix Australia Pty Ltd. These contaminated sites include former gasworks, tanneries, wharfs, and military facilities with some sites located in desirable areas.

Well known brownfied sites in Australia include the Docklands area in Melbourne and the Sydney Olympic site in Homebush. Additional sites targeted for development at this time include the former Wesfarmers fertilizer plant site at Bayswater, WA, the Rhodes Peninsula area (adjacent to the Sydney Olympic site), the ADI site at St. Marys, west of Sydney, and the Green Square area, south of the Sydney CBD. These former industrial sites are being considered for residential and/or commercial development.

As with all brownfield sites, projects in Australia proceed through a familiar path of investigation and remediation. The former Australian Gas Light Plant in Sydney developed by Bovis Lend Lease (30 The Bond) is a typical example of the process. Soil and groundwater contamination from the former gas plant was remediated by excavation, installation of a groundwater barrier, and vapor suppression.

Each Australian state has its own somewhat unique approach to the cleanup of contaminated sites. This includes funding for the remediation of contaminated properties and establishing a publically available register of all confirmed contaminated properties.

Australia attempts to facilitate the development of brownfield sites through the planning and approval process and community engagement. The government also establishes clear requirements which to some degree takes the uncertainty out of the process, although any remediation project may not go entirely according to plan.

The Australian government initially addressed contaminated sites in 1999 through the National Environmental Protection Council under the National Environmental Protection Measure (Assessment of site contamination). This program is designed to provide, to the extent possible, consistent management practices for all stakeholders involved in the development of brownfield sites. This approach is in alignment with programs in the U.S. and the EU. Certainty in the regulatory process can facilitate the development of brownfield sites.

Australia is also moving beyond the mere remediation of brownfield sites and promoting sustainable development practices in urban redevelopment. This is part of a "global movement" to incorporate social factors into urban redevelopment. (http://www.infolink.com.au/n/Overview-of-brownfield-redevelopment-in-Australia-n757503)

Incentives

OVERVIEW

The most powerful means to promote the remediation and development of brownfield sites are financial incentives aligned with public policy. While regulations can ensure that brownfield sites are cleaned up, it is the incentives which will truly drive asset conversion into viable reuse of the sites.

We will examine primarily U.S. and European incentives in the remediation and reuse of brownfield sites and the positive impact these incentives, coupled with a strong regulatory and public policy framework, have had on transforming brownfields into productive developments.

U.S. federal regulatory programs, both CERCLA and RCRA, have been effective as a vehicle for the cleanup and redevelopment of brownfield sites. The ability to combine a regulatory framework and incentives appears to be the most effective in promoting redevelopment.

Moreover, there is a relatively recent nexus between brownfields redevelopment and job creation driven by the need to jump start economic growth resulting from the U.S. and global "economic meltdown" of late 2008. In response to the meltdown and the need to jump start economic growth in the U.S., the $789 billion "American Recovery and Reinvestment Act," affectionately known as the "stimulus bill," was signed on February 17, 2009.

The stimulus bill provides several hundred million dollars in direct funding for brownfields-related efforts, including $100 million for the assessment and cleanup of brownfields, $200 million for cleanup of leaking underground storage tanks, and $600 million for Superfund sites. The bill also provides indirect funding that could be used for remediation or redevelopment of brownfields in some situations. This includes $6 billion for water quality-related construction and improvement programs, $1 billion in community development block grants, and $3.2 billion for energy efficiency and reduction of carbon emissions, as well as other funding. The bill also includes billions of dollars in tax credits and government bonds.

The USEPA is deploying these stimulus bill funds by encouraging state and local governments and the private sector to redevelop brownfield sites. The recent USEPA "American Recovery and Reinvestment Brownfields Job Training Grants" is expected to promote not only the remediation of brownfield sites but the creation of "green jobs."

So what are green jobs? A "green-collar" worker is one who is employed in the environmental sectors of the economy. What has happened is that the definition of environmental sectors has expanded to include sustainability initiatives such as energy, water, and climate change/carbon with a revisit of brownfield sites. There appears to be a case by experts (www.greenforall.org) that a push toward a green economy will create millions of new jobs and help the U.S. economy recover while improving the environment.

This relatively recent development in the economic landscape, the notion of "green jobs," ties the surge in new "clean technologies" such as solar, wind, green building, etc. with job creation. The investment in brownfield sites, perhaps one of the earliest efforts of the "green movement," is also tapped to create new jobs as part of the overall push to create economic growth in the U.S.

The recent publication of "The Green Collar Economy" by Van Jones (founder of the organization "Green for All") best captures the thesis that we are moving to an economy focused on green technologies which will create new companies, products, services, and jobs. In March 2009 Van Jones was brought into the new U.S.

administration as a "Green Jobs Advisor" to work with U.S. federal agencies and departments to advance energy and climate initiatives and, of course, create green jobs.

Before we move onto what we more traditionally view as incentives for brownfields redevelopment, one last word on green job creation in the U.S. (at least of this chapter). Sustainable South Bronx (SSBx) is a very progressive organization that illustrates the nexus between environmental justice initiatives, brownfields redevelopment, and green job creation. I should first define environmental justice which refers to "inequitable environmental burdens born by groups such as racial minorities, women, residents of economically disadvantaged areas, or residents of developing countries." SSBx, founded in 2001 by Majora Carter, a long time resident of the South Bronx, has as its mission to create "environmental justice solutions through innovative, economically sustainable projects informed by community needs." SSBx addresses "land-use, energy, transportation, water & waste policy, and education to advance the environmental and economic rebirth of the South Bronx, and inspire solutions in areas like it across the nation and around the world."

I call attention to this to illustrate the point that brownfields are no longer just about reuse or building green; brownfields can become the catalyst to address issues such as environmental justice and green job creation.

Silo thinking no longer captures the most value from a brownfield site.

U.S. FEDERAL PROGRAMS

U.S. federal programs addressing brownfield sites are CERCLA (Superfund) and RCRA. Simply, CERCLA is designed to address abandoned contaminated sites and the RCRA brownfields programs is designed to address contaminated sites that will likely have some RCRA compliance requirement as part of any cleanup and reuse efforts.

An excellent overview of U.S. federal brownfields programs is available from the "Brownfields Federal Programs Guide" (USEPA August 2005, EPA-560-F-05-230). This guide is designed to be a technical resource and handbook to funding for U.S. brownfield projects. While I will focus on incentives and funding from USEPA,

CERCLA, and RCRA programs, the guide highlights finding opportunities with a diverse group of U.S. federal agencies from the U.S. Department of Agriculture to the U.S. Federal Housing Finance Board.

An overview of the CERCLA and RCRA brownfield programs will provide a sense for the landscape of funding and incentives. This overview is not meant to be exhaustive, but is instead intended to in effect provide a feel for the federal programs. A detailed analysis of relevant federal regulations which may apply to a particular property is best addressed by an environmental attorney.

CERCLA

The USEPA provides several incentive programs for the remediation of brownfield sites (http://www.epa.gov/brownfields). The USEPA "Brownfields Economic Redevelopment Initiative" was created to encourage states, communities, and other stakeholders to remediate and reuse brownfield sites. The initiative also included a forward looking provision to fund pilot projects to develop job training (green jobs) and address environmental justice issues (USEPA, Solid Waste and Emergency Response. EPA 500-F-97-155 April 1997).

Former President Clinton signed the "Taxpayer Relief Act" in August 1997 which included a tax incentive to promote the remediation and reuse of brownfield sites (HR 2014/PL 105-34). This program was essentially an extension of the Clinton administration's "Brownfields National Partnership Action Agenda" in May 1997.

This partnership essentially provided a more comprehensive approach to the remediation and reuse of brownfield sites through the coordination of 15 separate U.S. federal agencies.

With regard to U.S. federal tax law, "those expenditures that increase the value of, extend the useful life of a property, or that adapt the property to a different use, be capitalized; and, if the property is depreciable, that they are depreciated over the life of the property."

In plain English this means that the full cost can't be deducted from income in the year the expenditure occurs. The Internal Revenue Service (IRS) in 1994 ruled that certain costs incurred to remediate soil and groundwater could be deducted as a business expense in the same year. There were several problems with this: only the taxpayer who contaminated the land could deduct the expenses; it did not address any cleanup costs incurred by the purchaser of the contaminated property; and it was unclear if the expenses were deductable in the year incurred or could be capitalized.

These issues, while not insurmountable, apparently created enough of a problem to be an obstacle to remediating brownfield sites and returning them to productive use. The dilemma was that if you owned contaminated property and cleaned it up you could sell it at market value. On the other hand, if you were a prospective purchaser of a contaminated site, you had to buy it at "distressed value" and capitalize the remediation costs. In addition, if as an owner you wanted to upgrade the use of the site, you were not able to fully deduct your costs to remediate the site.

Under the Brownfields Tax Incentive, environmental cleanup costs for properties in targeted areas are fully deductible in the year in which they are incurred, rather than having to be capitalized. At the time the $1.5 billion incentive was expected to leverage $6.0 billion in private investment and return an estimated 14,000 brownfields to productive use. The Brownfields Tax Incentive was a smart federal law that promoted the remediation, reuse, and sale of brownfield sites.

This tax incentive applies to properties that meet specified land use, geographic, and contamination requirements. The requirements for eligibility are:

- The property must be held by the taxpayer incurring the eligible expenses for use in a trade or business or for the production of income, or the property must be properly included in the taxpayer's inventory.

- Hazardous substances must be present or potentially present on the property.

- The property must be located in one of the following areas:
  - "EPA Brownfields Pilot areas designated prior to February 1997;
  - Census tracts where 20 percent or more of the population is below the poverty level;
  - Census tracts that have a population under 2000, have 75 percent or more of their land zoned for industrial or commercial use, and are adjacent to one or more census tracts with a poverty rate of 20 percent or more; and

– Any Empowerment Zone or Enterprise Community (and any supplemental zone designated on December 21, 1994)."

The tax incentive applies to both urban and rural sites and the owner (taxpayer) is required to certify that the site qualifies for the incentive. Finally, sites that are listed on the UESEPA National priorities List (NPL) are excluded.

In 2001, the Small Business Liability Relief and Brownfields Revitalization Act amended CERCLA by providing funds to assess and clean up brownfields, providing financial assistance to protect human health and the environment, and either promote economic development or enable the creation of, preservation of, or addition to parks, greenways, undeveloped property, other recreational property, or other property used for nonprofit purposes (http://www.epa.gov/swerosps/bf/sblrbra.htm).

H.R. 2869 was introduced in the House in September 2001 and essentially combined two bills which amended CERCLA. H.R. 2869 incorporated S. 350, the "Brownfields Revitalization and Environmental Restoration Act of 2001" with H.R. 1831, the "Small Business Liability Provision Act." As previously introduced, S. 350 contained three "titles" dealing with the following:

- "Title I codified and expanded EPA's current brownfields program by authorizing funding for assessment and cleanup of brownfields properties;
- Title II exempted from Superfund liability contiguous property owners, prospective purchasers, and clarified appropriate inquiry for innocent landowners; and
- Title III authorized funding for state response programs and limited EPA's Superfund enforcement authority at sites cleaned up under a state response program."

H.R. 2869 also incorporates H.R. 1831, the "Small Business Liability Protection Act." The Small Business Liability Protection Act exempted "de micromis" contributors of hazardous substances along with household, small businesses, and nonprofit entities which generated solid waste from CERCLA liability. Basically the bill provides for expedited settlements for those with a limited ability to pay.

Title II provided for brownfields funding through Section 211. The key elements of Title II were to:

- Authorize up to $200 million per year for brownfields assessment and cleanup to carry out new section 104(k). Included $50 million per year or 25 percent of amount appropriated to carry out section 104(k) of the bill, for brownfields with petroleum contamination.
- Define brownfield sites: real property, the expansion, redevelopment, or reuse of which may be complicated by the presence or potential presence of a hazardous substance, pollutant, or contaminant.
- Purposes of section 104(k):

    1. Land contaminated by petroleum or petroleum products;
    2. Land contaminated by a controlled substance as defined in the Controlled Substances Act (21 U.S.C. 802); and
    3. Mine-scarred land.

    Exclusions:
    1. Subject to a planned or ongoing CERCLA removal action;
    2. Listed or proposed for listing on the National Priorities List;
    3. Subject to a unilateral administrative order, court order, administrative order on consent, or consent decree under CERCLA;
    4. Subject of a unilateral administrative order, court order, administrative order on consent, consent decree, or permit under RCRA, CWA, TSCA, or SWDA;
    5. Subject to corrective action under RCRA 3004(u) or 3008(h) to which a corrective action permit or order has been issued or modified requiring the implementation of corrective measures;
    6. Land disposal units with closure notification submitted and closure plan or permit;
    7. Subject to the jurisdiction, custody, or control of federal government;
    8. With PCB contamination subject to remediation under TSCA; and
    9. Which have received assistance from the Leaking Underground Storage Tank for a response activity.
- Provide authority to include some otherwise excluded sites on a site-by-site basis.

- Eligible entities for brownfields funding include states, tribes, local governments, land clearance authorities, regional councils, redevelopment agencies, and other quasi-governmental entities created by states or local governments.
- Imposes significant restrictions on charging administrative costs to grants.

With regards to incentives and funding, the act provided for the following:

- Brownfields site characterization and assessment
  1. Authorizes grants of up to $200,000 per site to eligible entities to inventory, characterize, assess, and conduct planning at brownfield sites;
  2. Authorizes targeted site assessments at brownfield sites; and
  3. National Contingency Plan (NCP) requirements may be imposed only when relevant and appropriate to the program.
- Brownfields remediation:
  1. Authorizes grants of up to $1 million to eligible entities to capitalize revolving loan funds to clean up brownfields;
  2. Authorizes grants of up to $200,000 per site to eligible entities or nonprofit organizations to cleanup brownfields owned by the grant recipient;
  3. Grants generally require a 20 percent match;
  4. Construction, alteration, and repair work funded all or in part with grant funds is subject to Davis Bacon Act; and
  5. NCP requirements may be imposed only when relevant and appropriate to the program.
- Brownfields program:
  1. Establishes program to provide training, research and technical assistance to facilitate brownfields assessment and cleanup; and
  2. Limited to 15 percent of amount appropriated to carry out section 104(k).

While not directly related to funding or incentives there are other provisions of the Act worth noting as related to liability.

Subtitle B provided for clarification and exemptions for landowners and operators. Specifically, the following elements were included:

- Exempts from owner or operator, liability persons that own land contaminated solely by a release from contiguous, or similarly situated property owned by someone else, if the person:
  1. Did not cause or contribute to the release or threatened release;
  2. Is not potentially liable or affiliated with any other person potentially liable;
  3. Exercises appropriate care in respect to the release;
  4. Provides full cooperation, assistance, and access to persons authorized to undertake the response action and natural resource restoration;
  5. Complies with all land use controls and does not impede the performance of any institutional controls;
  6. Complies with all information requests;
  7. Provides all the legally required notices regarding releases of hazardous substances; and
  8. Conducted all appropriate inquiry at time of purchase and did not know or have reason to know of contamination.
- Prospective purchasers and windfall liens
- Exempts bona fide prospective purchasers (and their tenants) from owner or operator liability so long as the person does not impede the performance of a response action or natural resource restoration. The definition of a bona fide prospective purchaser is one where: "all disposal took place before the date of purchase; person made all appropriate inquiry; person exercises appropriate care with respect to any release; provides full cooperation, assistance, and access to persons authorized to undertake response actions or natural resource restoration; complies with land use restrictions and does not impede performance of institutional controls; complies with all information requests; provides all legally required notices regarding releases of hazardous substances; person is not potentially liable or affiliated with any other person potentially liable; and provides the U.S. with a lien on the property if the U.S. has unrecovered response costs and the response action increases the fair market value of the facility.

The act also provided for greater clarification with regard to "Innocent Landowners." Key elements of this definition are as follows:

- Clarified what actions landowners must take to satisfy the "all appropriate inquiries" requirement of the defense;

- Directed the USEPA to promulgate within two years regulations establishing standards and practices for satisfying the all appropriate inquiries requirements;

- Until the USEPA issued the required regulations, two standards applied depending on the date the property was purchased (note the ASTM standard below):
  1. "Prior to May 31, 1997—A court shall consider specialized knowledge of the defendant, relationship of purchase price to value of uncontaminated property, commonly known information, obviousness of contamination, ability of defendant to detect contamination by appropriate inspection."
  2. After May 31, 1997—ASTM "Standard Practice for Environmental Site Assessment: Phase 1 Environmental Site Assessment Process."

- In the case of a facility purchased for residential use by a person who is not a government or commercial entity, a facility inspection and a title search satisfy the appropriate inquiry requirement.

The act also provided for state funding for response programs which will be described in greater detail in the following section.

- Authorized $50 million per year for grants to assist states and tribes in the development of state response programs.

- A state may be awarded funds if it is a party to a memorandum of agreement with USEPA for its voluntary response program, or if the state includes, or is working toward including, the following elements in its program:
  1. Timely survey and inventory of brownfield sites;
  2. Oversight and enforcement authorities to ensure protection of human health and environment;
  3. Meaningful public participation; and
  4. Mechanism for approval of a cleanup plan and certification that response is complete.

Also included was a restriction of federal administrative or judicial enforcement action under CERCLA 106(a) or cost recovery actions under CERCLA 107(a) at any "eligible response site" at which there is a "release, or threatened release, of a CERCLA-covered substance and at which a person is conducting a response in compliance with a state program that specifically governs response actions for protection of human health and the environment." The definition of an "eligible response site" is listed as: certain LUST sites; certain sites covered by RCRA, CWA, TSCA, or SDWA excluded from the definition of a brownfield site, if, as determined on a site-by-site basis, findings are made that not taking enforcement will still limitations on enforcement are appropriate and will (1) protect public health and the environment and (2) promote economic development or open space." Sites not considered as "eligible response sites" were facilities at which federal preliminary assessments or site inspections are conducted and are qualified for listing on the NPL and facilities determined to warrant particular consideration, as identified by regulation (such as threats to a drinking water aquifer or a sensitive ecosystem). (http://www.epa.gov/swerosps/bf/html-doc/2869sum.htm)

The "Brownfields Tax Incentive" was signed into law in October 2008. The tax incentive known as H.R. 1424 (specifically section 318) provides for the ability to expense environmental remediation costs extending the current brownfield tax incentive until December 31, 2009, and is effective for expenditures paid or incurred after December 31, 2007. The Brownfields Tax Incentive allows for taxpayers to receive a current federal income tax deduction for certain qualifying remediation costs that would otherwise by subject to capitalization. (http://www.epa.gov/brownfields)

The Brownfields Assessment, Revolving Loan Fund, and Cleanup Guidelines outlined in 2009 are designed to address the need to remediate sites contaminated by hydrocarbons, hazardous substances, and other contaminants. Funding is provided for the following:

- Brownfields Assessment Grants (each funded up to $200,000 over three years);
- Brownfields Revolving Loan Fund (RLF) Grants (each funded up to $1,000,000 over five years); and
- Brownfields Cleanup Grants (each funded up to $200,000 over three years).

Additional information regarding the Grant, Revolving Loan, and Cleanup Grant guidelines is available at the USEPA Brownfields and Land Revitalization Web page (http://www.epa.gov/brownfields).

As previously mentioned, the USEPA also provides funding for Brownfield Sustainability Pilots. USEPA provides technical assistance for the sustainable assessment, cleanup, and redevelopment of community brownfield projects. Assistance supports activities such as the reuse and recycling of construction and demolition materials, green building and infrastructure design, energy efficiency, water conservation, renewable energy development, and native landscaping (http://www.epa.gov/brownfields).

RCRA

In addition to CERCLA, the USEPA RCRA program also provides funding for brownfields redevelopment projects. Under the RCRA brownfields program, a brownfields site is a "facility or portion of a facility where development has been delayed due to real or perceived concerns that it is contaminated with hazardous waste and will need to comply with RCRA and/or go through the RCRA corrective action process to achieve cleanup." Unlike the USEPA CERCLA program which is designed to address abandoned sites, the RCRA brownfields program generally works with sites that have a current property owner or operator. Sites covered under the RCRA brownfields program might not be currently regulated under RCRA but the land may be contaminated with hazardous wastes that may trigger RCRA requirements when the property is remediated.

The RCRA brownfields program includes a pilot program, where USEPA partners with states, communities, and facilities to highlight programs where corrective action efforts are expedited. More importantly, the pilot program was developed to move underutilized or inactive facilities to cleanup and productive use (http://www.epa.gov/swerosps/rcrabf/pdf/introfs.pdf).

The RCRA program also includes an outreach program to communities to convey lessons learned from the pilot projects to increase access to innovative approaches, tools, and resources.

The Targeted Site Efforts (TSE) aspect of the RCRA brownfields pilot program is designed for facilities where every effort would be required to move toward and through site cleanup and reuse. This is essentially the implementation of a targeted action to jump start cleanup and 'asset conversion," moving the site back to productive use. The TSE is an interesting approach as it recognizes that sometimes regulatory programs need a push in getting sites through the program without compromising the integrity of the regulatory framework.

Finally, the RCRA program in general has undergone "Cleanup Reforms" designed to create a more flexible and faster RCRA program that can interface with state voluntary cleanup programs to achieve the mutual goal of transforming brownfield sites into productive redevelopment opportunities.

In summary the RCRA brownfields program (in a similar fashion as the CERCLA brownfields program) provides for site characterization and assessment grants, grants and loans for remediation, job training grants, revolving loan fund grants, and training, research, and technical assistance grants. Funding is available to local governments, land clearance authorities, or quasi-governmental organizations, regional councils, redevelopment agencies, states, non-profit organizations, entities created by state legislation, Indian Tribes, and Alaska Native Regional Corporation (http://www.epa.gov/swerosps/rcrabf/pdf/bfunding.pdf).

Other U.S. Federal Programs

While CERCLA and RCRA are the "backbone" of U.S. federal regulatory programs there are several other U.S. federal programs that provide incentives for brownfields redevelopment. These federal programs provide an interrelated web of funding opportunities that, in some cases, can ensure a project overcomes funding hurdles. While it may appear daunting to navigate through these programs, the time invested is worthwhile. Couple these federal programs with state programs and numerous resources are available to the private and public sectors.

The USEPA Brownfields Federal Programs Guide, published in 2005, provides a summary of nearly two dozen federal programs that can help communities assess, cleanup, and reuse brownfields. Typically, programs through USEPA and HUD focus explicitly on brownfields, but communities applying some creativity can benefit from many other federal programs that can provide technical and financial support to brownfields cleanup and redevelopment (http://www.epa.gov/brownfields/partners/2005_fpg.pdf).

These "other" federal programs include:

- New Markets Tax Credit established by the Community Renewal Act of 2000 under the U.S. Department of Treasury to stimulate the economies of low-income communities (http://www.epa.gov/brownfields/partners/2005_fpg.pdf).
- Low Income Housing Tax Credits is available to developers and investors with affordable housing developments (http://www.epa.gov/brownfields/partners/2005_fpg.pdf).
- Federal Historic Preservation Tax Incentives established under the National Park Service and Internal Revenue Service for the rehabilitation of historic buildings and properties (http://www.epa.gov/brownfields/partners/2005_fpg.pdf).
- Community Reinvestment Act of 1977 encourages a cooperative approach amongst federal partners, lenders, and purchasers to ease fears of financial liabilities and regulatory burdens. Lenders subject to the CRA can obtain credits for lending to brownfield revitalization projects in low and mid-income areas (http://www.epa.gov/brownfields/partners/2005_fpg.pdf).

U.S. STATE PROGRAMS

U.S. federal CERCLA and RCRA programs coupled with U.S. state programs have been effective in promoting the reuse of brownfield sites through increased funding and relative "ease" in navigating various regulatory programs. Brownfields programs are in place in nearly all 50 states and include one or more of the following (http://www.epa.gov/brownfields/partners/finan_brownfields_epa_print.pdf):

- Tax incentives through credits, abatements, and forgiveness;
- Direct financing through the use of loans and grants; and
- Offsets to brownfield financing through technical assistance, process facilitation, project support, tools, and/or flexible cleanup programs.

State programs are typically focused on achieving one or more of the following objectives:

- Reduce risk by providing loan guarantees, insurance, or clarification of legal issues to promote access to capital;
- Reduce cost for the borrower to finance redevelopment by subsidizing loan carrying costs or providing reduced loan underwriting to, in some cases, promote the development of smaller projects;
- Reduce financial risk by providing incentives to improve project cash flow; and
- Provide direct financial assistance, including loans or grants, to defray remediation costs and support site redevelopment.

With wide variation from state to state these program typically fall into three major categories:

- Tax incentives linked to site reuse;
- Direct financial assistance; and
- Offsets to brownfields financing needs, including technical assistance, process facilitation, and project support.

State programs are relatively common and very diverse in approaches. Despite the diversity of approaches to promoting brownfields remediation and reuse, states are reaping the benefits of these programs. While the economic downturn of late 2008 has significantly impacted development projects and created new brownfield sites resulting from shuttered commercial and industrial facilities, one can expect a resurgence in development once the U.S. economy resumes growth in late 2009 or early 2010.

Public financing is often essential to encourage the remediation and redevelopment of brownfield sites. Without public financing and government-backed incentives, most brownfield sites would remain idle and contaminated. Additional costs associated with redevelopment of a brownfield site include demolition, environmental assessment activities, and remediation actions to reduce risks to human health and the environment and address liability concerns. As a result of these additional costs, government incentive programs are often necessary to make redevelopment of brownfield sites competitive with greenfields sites.

States offer a variety of incentive programs with the objectives to: reduce lender risk by providing off-setting incentives such as loan guarantees, insurance, or property-specific legal clarifications to make capital more readily available; reduce borrower's cost of financing by subsidizing loan carrying costs or by providing assistance that reduces loan underwriting and documentation expenses; ease a purchaser's financial risk with incentives that improve project cashflow such as tax credits or abatements; provide direct financial assistance, including loans or grants to pay for assessment and cleanup activities and support broader redevelopment needs. Three general categories of state incentives include tax incentives linked to site reuse; direct financing assistance; and offsets to brownfields financing needs, including technical assistance, process facilitation, and project support.

A key feature of many state programs is not only to provide incentives for cleanup of properties, but for the return of these properties to productive use. A number of states are focusing incentives based on property type (e.g., dry cleaners, manufacturing sites, etc). Many states are also encouraging local governments to use traditional public financing tools such as tax increment finance (TIF) for brownfield sites. There has been considerable creativity in promoting brownfields cleanup and redevelopment, including tax forgiveness and expedited tax foreclosure for new owners performing cleanup under state VCP, cleanup and redevelopment tax credits, including site preparation credits and environmental remediation insurance tax credits. In addition, programs provide:

- Low interest loans to redevelopment agencies and nonprofits to purchase contractor liens, tax certificates, and other property claims;
- Direct grants for phase II assessments available to companies and developers; and
- Establishment of dedicated funds to address cleanup and redevelopment of operating or abandoned dry cleaner sites.

These incentive programs are creating jobs, leveraging private investment for the creation of affordable housing, improving tax revenues, reducing risks to human health and the environment, and encouraging revitalization of declining neighborhoods.

Tax incentives are typically used to offset cleanup costs or provide a buffer against increases in property value that result in tax increases before site preparation costs are paid off. Tax incentive support typically includes deferral of increased property taxes, remediation tax credits, cancellation of back taxes, rebates of sales taxes, environmental insurance tax credit, business tax offset, tax "bonus refund" pegged to job creation, and state historic rehabilitation tax credits.

Direct financing such as grants and loans provide capital needed for financing brownfields cleanup and reuse. Some programs help finance specific parts of the project, while others increase the lender's comfort by limiting the risk of potential losses or loan defaults. These programs can also potentially ease the borrower's cash flow. Nearly half of states offer direct financing including forgivable remediation loans, low interest loans for contractor/lien purchases, low-interest cleanup loans, remediation grant funds, state revolving loan or redevelopment funds, and loan guarantees. Some states earmark funds directly for specific brownfield initiatives such as phase I and II assessments. States have been able to utilize USEPA funded Clean Water State Revolving Loan Funds to provide loans for brownfield sites which result in improved water quality, and Federal Highway Administrative funds have been used to support brownfields-related transportation improvement projects.

Offset programs consist of technical assistance, process facilitation, and project support, including insurance subsidies, brownfield redevelopment authorities, and the programs focus on agricultural-related contaminants. These programs reduce the necessary level of capital investment, making these projects more economically viable.

Approximately 75 percent of the states have programs in place to facilitate financing with minimal cash outlay, including facilitation of property transfer of orphan sites to new purchasers as part of an agreement to cleanup the property, programs that allow use of institutional controls to contain contamination rather than cleanup to background or residential levels, encourage enrollment into VCUP programs which can provide information on how brownfield programs can facilitate access to other financing tools, provide finality to the cleanup via liability relief, provide a level of comfort due to state oversight, and provide notices of "no further action." States are promoting brownfields reuse projects by making slight alterations to their existing economic development finance programs, i.e., recognizing site assessment and remediation needs as legitimate project development activities within the context of their common financial assistance incentives.

As late as 2007 some of the benefits from state brownfield programs (according to the USEPA http://www.epa.gov/

brownfields/partners/finan_brownfields_epa_print.pdf)
include the following:

- Wisconsin benefitted from over 4000 new jobs attributed from
  88 brownfield projects;
- Minnesota estimated that its brownfields program leveraged
  nearly $1 billion in private sector investment to create 5700
  housing units; and
- Florida created 3274 jobs and $172 million in new investment
  from brownfields redevelopment.

  A sampling of the diversity of the programs is provided below:

- Tax Incentives—as of 2007 approximately 23 states have tax incentive components where tax credits, tax abatements, or forgiveness can increase project cash flow. This improved cash
  flow can attract lenders thereby increasing the probability of
  project development. The state tax incentives to support brown-
  fields financing include:
  - Deferral of increased property taxes—Connecticut and
    Texas;
  - Remediation tax credits—Illinois, Ohio, and Wisconsin;
  - Cancellation of back taxes—Wisconsin and Massachusetts;
  - Rebates of sales taxes to offset cleanup costs—New Jersey;
  - Tax incentives to provide financial flexibility—Missouri;
  - Environmental insurance tax credit—Michigan; and
  - Tax "bonus refund" pegged to job creation—Florida.
- Direct financing—states have begun to increase their focus on
  direct financing (although I am convinced this is currently on
  hold as states are strapped for funding) to promote brownfields
  redevelopment. Almost 50 percent of the states provide some
  degree of brownfields financing including the following:
  - Forgivable remediation loans for expanded petroleum
    sites—Indiana;
  - Low interest loans for contractor/tax lien purchases—
    Florida;
  - Low interest cleanup loans—Delaware, Indiana, and Wisconsin;
  - Remediation grant funds—New Jersey and Minnesota; and

- Revolving loan or redevelopment funds—Indiana, Michigan, Wisconsin, and Massachusetts.
- Offsets to brownfields financing—such as technical assistance, process facilitation, and project support are provided as part of programs in:
  - Insurance subsidies—Massachusetts and Wisconsin;
  - Brownfield Redevelopment Authorities—Michigan; and
  - Focus on agricultural-related contaminants—Kansas.

THE EU

The EU "Regeneration of European Sites in Cities and Urban Environments" or RESCUE program provides a robust framework for the redevelopment of brownfield sites. The RESCUE "Best Practice Guidance for Sustainable Brownfields Regeneration" published in May 2005 outlines how the EU and member countries can meet the challenges of a sustainable approach to development, including the identification and evaluation of sustainable brownfield redevelopment practices (http://www.rescue-europe.com/download/reports/RESCUE%20Manual.pdf). I will get into the EU RESCUE program later as part of the discussion of "frameworks" for moving brownfield sites through green development.

The EU provides an array of incentives through several funding channels. A brief summary of financial incentives and resources to guide through the various programs is provided below.

- EU level financial incentives summary (ERDF, ESF, URBAN, INTERREG, PHARE, ISPA); Public Credit (European Bank for Reconstruction and Development); Research (LIFE, 6th Framework Programme).
- EU level legal incentives summary—Bans (Landfill Directive); Obligations (UN/ECE Convention on Access to Information, Landfill Directive); Guidelines/Principles (Handbook on Environmental Assessment of Regional Development Plans and EU Structural Fund Programmes).
- National level subsidies and tax credits, release of liability, national funds, preferential loans, grants, tax exemptions, VAT reductions.

- National level legal incentives—Landfill restrictions, protection zones, environmental protection laws, comprehensive brown-field strategies, and sustainable community plans.

A comprehensive overview of EU financial incentives is pro-vided in the EU Regional Policy Conference Proceedings, Regions for Economic Change—Sharing Excellence, February 2008 (http://ec.europa.eu/regional_policy/conferences/competitive-ness/doc/presentations/workshop3b/groenendijk_3b.ppt). Similar to the U.S. programs, the EU incentives consist of the following financial "tools":

- Cash grants;
- Loans;
- Tax incentives;
- Risk insurance and relief;
- Liability relief;
- Capital attraction incentives;
- Planning and land assembly assistance;
- Tax increment financing;
- Revolving loan funds;
- Benefit sharing;
- Development charges;
- Development gains taxes and planning gain supplement;
- Integrated contracts; and
- Portfolio development.

Frameworks for Greening Brownfields

OVERVIEW

It is convenient to have a framework in mind when working through the investigation and redevelopment of a brownfields site. While elements of the process are relatively straightforward, such as the onsite investigation, the overall process of integrating all of the phases tends to be unique for each site and more fluid. I am convinced that no one framework would fit every site redevelopment scenario. Instead, there are a few convenient frameworks to assist in thinking through the elements of the redevelopment process. A few of these will be presented and a simplified framework will be discussed.

Colorado Brownfields Handbook

Let's start with "The Colorado Brownfields Handbook" as it provides a simple "roadmap" for site redevelopment. According to the Colorado Handbook the role of local government is to facilitate the redevelopment process by providing:

- "Visioning—Recognize a community or economic need;

79

- Formulating reuse scenarios;
- Evaluating business opportunity, financial viability, economic impacts, and environmental conditions for the reuse of a property;
- Transaction—Resolving risk management issues to facilitate property title transfer (if necessary); and
- Implementing redevelopment—Conduct environmental remediation, construction, and renovation steps, and ultimately sell the property."

Although the Colorado handbook obviously has a local perspective it lays out a visioning and facilitation exercise which can be universally applied. The key to the greening of brownfield sites is being proactive and engaging all stakeholders. This is seldom done well, but when it is, the project team benefits greatly.

The visioning process must include the community. The rationale is simple—there are a number of broad social, economic, and land use issues that are uniquely important to the community. Such issues include; the overall direction of the community, addressing a workforce housing shortage, a desire for additional open space preservation, or the creation of additional commercial projects. In some cases brownfields redevelopment is the catalyst or at least the first step in moving along a strategic economic growth path for a community.

In thinking about a "reuse scenario" there are several key factors to be built into the process (http://www.cdphe.state.co.us/HM/bfhandbook.htm). They include:

- Property condition. An assessment of the property condition is essential and includes an analysis of the site, building structure, mechanical and electrical systems, safety issues, recommended building code compliance reviews, building interior, and environmental conditions.
- Community impact. Evaluate compatibility with community goals, planning and zoning, public safety issues, and value impacts on surrounding property values.
- Context. Evaluate the relationship with surrounding properties, "area vitality," and stakeholders.
- Opportunity. Identify economic assets, economic development opportunities, reuse scenarios, and resulting economic impacts.

- Implementation Strategies. Identify key stakeholders, partnering opportunities, and funding mechanisms.

The overall assessment needs to answer the following:

- Is the site useable or marketable as-is?
- What needs to be done to make the site useable/marketable?
- Is there a public benefit from reuse?
- What are the hurdles to reuse?
- What strategies could be pursued to facilitate reuse?

The Colorado brownfields guide makes a key point. If the site is not immediately well suited for redevelopment, at the very least it can be *positioned* for redevelopment. The redevelopment process can take a considerable amount of time. An early start can pay off.

For those communities that take the lead on the redevelopment of brownfield sites or at the very least are very actively engaged, the Colorado Brownfields Guide highlights the following actions (from the Colorado Brownfields Manual, quoted and paraphrased below).

- "Education. Educating property owners, developers, businesses, lenders, and city/county departments to overcome misperceptions and build support for local projects. These stakeholders are often uninformed about brownfield solutions and fear potential liability and reduced property marketability after redevelopment. Conveying information about risk-based cleanup approaches, cost-effective engineering solutions, liability management options, and available funding programs is important in generating interest in brownfields redevelopment.

- Integration of community priorities. Cleanup and reuse can address multiple community concerns, such as lack of space for business and housing, property maintenance and improvement issues, vandalism, public safety concerns, and declining tax bases.

- Coordination of intra-governmental relations. Because brownfields redevelopment is a land use and development activity, traditionally independent government departments may have a common interest in a project. These departments might include: economic development, planning, public works, environmental/solid waste, housing, public safety, engineering, transportation, health and human services, and legal. Use a team approach to explicitly involve appropriate departments.

- Coordination of intergovernmental relations. It is essential to ensure communication and cooperation between city, county, and state contacts for securing project approvals, funding assistance, and closing regulatory environmental issues. Investors ascribe economic value to the regulatory benefits provided by voluntary cleanup programs and the most effective jurisdictions work closely with state and federal environmental regulators.

- Coordination of various stakeholder groups. Brownfields redevelopment can be public sector driven, private sector driven, or a combined effort depending on the project. As with other land development activity, identifying appropriate parties and managing relationships, including with the community, is essential to a successful project.

- Provide an information clearinghouse. It may be necessary or desirable to engage the interest of businesses and developers seeking locations with brownfields opportunities. This may simply entail broadening the vision of economic and business development services that many jurisdictions already provide. Nationally, some municipalities maintain an inventory of brownfield sites for planning purposes and to prioritize investment opportunities, while others see inventories as too costly or stigmatizing certain properties.

- Coordination and/or provision of funding. There are many ways to enhance project viability and address brownfields issues. This may include grants and low-interest loans to pay for environmental investigation, cleanup, and construction activities. Local efforts (such as economic development subsidies and tax incentives) can also increase the project feasibility. It may also entail identifying and packaging bank financing, outside governmental funding, and nonprofit sources of capital. In some instances, direct public investments in infrastructure, site acquisition, risk management, or other project-related outlay is warranted.

- Utilization of a coordinator. The numerous issues involved in brownfields redevelopment often make them too complex for any single person or agency to understand fully and direct experience with environmental regulations is often limited. It may be useful to designate a staff person, hire a consultant, engage a nonprofit organization or borrow a state or federal facilitator to implement brownfields activities."

Although I have not substantively discussed actual remediation technologies or the process (phase I, phase II and phase II activities) it is worth mentioning here that any brownfields redevelopment project should include the consideration of the use of environmental insurance. Insurance is a valuable tool in managing the financial uncertainty associated with the cleanup of a brownfields site.

The types of insurance available are:

- Cleanup cost-cap (stop-loss) coverage—Places a limit on the cleanup costs site redevelopers might have to incur.

- Pollution loss liability—This coverage applies to costs associated with any future required site cleanup including re-openers where regulation changes regarding a known condition necessitate additional, post-closure site work.

The bottom line for insurance is that it has the potential to create a more predictable investment scenario and has the potential to enhance project feasibility.

There are a few other frameworks and process decision trees that are worth highlighting.

Ontario Brownfields Toolbox

The Ontario Brownfields "Toolbox" provides a "brownfields decision tree" which lays out (rather simply and concisely) the process for redeveloping brownfield sites (http://www.aboutremediation.com/toolbox/faqs.asp). The Web site and toolbox were developed by aboutREMEDIATION (AR) and is Canada's "leading information resource on site remediation and brownfields redevelopment." The site launched in March 2001 (www.aboutremediation.com) and is owned and operated by the Ontario Centre for Environmental Technology Advancement (OCETA).

The site is any excellent resource for any brownfields redevelopment project.

Missouri Roadmap

The State of Missouri has also developed a simple and concise roadmap (http://www.stlrcga.org/documents/missouri_brown-field_guide.pdf/MISSOURI). The roadmap was developed as part of a St. Louis effort titled, "Turning Missouri Brownfields into Good Deals." The goal of the effort is to:

- "Promote one of St. Louis's most underutilized resources; and
- Demonstrate the viability of brownfield development opportunities given the litany of tools, programs, and technical assistance available in the St. Louis metropolitan region."

RESCUE Roadmap

Likely the most ambitious roadmap, framework, and process document I have seen is the EU "RESCUE—Best Practice Guidance for Sustainable Brownfield Regeneration (http://www.rescue-europe.com/index_mf.html)." RESCUE is, "Regeneration of European Sites in Cities and Urban Environments."

This is also very likely the best acronym for a brownfield sites guidance document I have seen to date.

The RESCUE "system" outlines a very comprehensive approach for the redevelopment of brownfield sites in the manual "Best Practice Guidance for Sustainable Brownfield Regeneration." The manual provides the following elements as part of the overall guidance:

- Definition of sustainable brownfield regeneration;
- Best practice examples in brownfield regeneration;
- "User tools" for the management of soil and contamination, the management of existing buildings and infrastructures, sustainable land use and urban design, sustainable planning processes and methods for citizen participation, the sustainable management of brownfield regeneration projects;
- Administrative tools and incentives for sustainable brownfield regeneration;
- Virtual Training Centre (VTC) which provides Web-based training resources for sustainable brownfield regeneration

Perhaps the most useful tool is the "Sustainability Assessment Tool (RESCUE-SAT)" which provides a methodology to promote the consideration of multiple project parameters and sometimes conflicting priorities to evaluate the viability of brownfields redevelopment projects.

In my opinion the most important aspect of the RESCUE effort and associated tools is that sustainability is well integrated into brownfields redevelopment. It is not viewed as an add-on but instead as an integral element of brownfields redevelopment. Moreover, historically, brownfields redevelopment focused on the technical aspects of the project (in particular the remediation). RESCUE brings the technical elements along with the "softer" elements of project development.

The RECUE process blends the economic, social, environmental and institutional elements of a project into view and attempts to lay out a viable framework for managing these project aspects.

The RESCUE initiative started by analyzing current practices of brownfields redevelopment in six industrial core regions of Europe. The analysis focused on how sustainability practices were embedded into brownfield projects.

The authors of the manual best captured the most important aspects of turning a brownfields site green. They state that:

"Sustainability cannot be defined generally for all brownfield regeneration projects since a land use, design, or methodology that proved to be suitable at one site is not necessarily appropriate for another site, another context, another time, or another mix of stakeholders with a different set of priorities. To strive for sustainability, brownfield regeneration projects need a holistic approach which comprises economic, environmental, social, and also institutional aspects."

They also highlight that brownfield regeneration requires:

- "Creativity and vision
- A well structured development process with clearly defined project milestones
- The involvement of a wide range of different experts' opinions in the development processes
- Good communication and citizen participation
- Comprehensive spatial strategies and collaboration on the regional level
- To be embedded into regional or local economic strategies that target to develop economic clusters and growth sectors

- Integration into the specific regional, local, and neighborhood contexts
- An integrated view on planning and remediation activities."

This is really the best summary I have seen of the process and considerations for the redevelopment of brownfield sites and how to integrate sustainability principles into projects. I encourage reading the manual. Figure 5.1 illustrates the RESCUE sustainability objectives.

I found the framework/decision tree most valuable when compared to the "decision chart" example provided in the manual for the Radbod, Germany site illustrated in Figure 5.2. The RESCUE framework makes the Radbod example relatively clear and transferable for use at other project sites.

RE-DESIGN-IT

A much simpler mental model for the greening of brownfield sites is illustrated in Figure 5.3. It is designed to highlight the key elements in greening a brownfields site and not intended to be comprehensive. The key elements of "RE-DESIGN-IT" are consideration of the remediation, design, and integration of the site into the community. It is also not meant to imply that the greening of a brownfields site is linear. It is most certainly not. However, the framework graphic is designed to illustrate that every project should consider all of the aspects highlighted and that the ultimate goal is to integrate the project into the community in a sustainable manner.

Creating an opportunity for "green living," the creation of new jobs (green jobs perhaps?), off grid energy, etc. is essential.

I am convinced we are in the process of "reinventing everything." Brownfields going green, along with all of the economic and social benefits, is part of this process. RE-DESIGN-IT is meant to capture that theme and promote movement toward a more sustainable world.

1. RESCUE SUSTAINABILITY OBJECTIVES				2. Responsibility for objective									
				Project Managers	Landowners	Developers	Planners	Policymakers	Regulators	Citizens	Contractors	Designers	Advisors
				3.	4.	5.	6.	7.	8.	9.	10.	11.	12.
13.	1	14.	To reduce negative environmental impacts on the site and on the neighbourhood including human health risks								X		
25.	2	26.	To minimise waste and maximise recycling and reuse of soil and debris								X		
37.	3	38.	To ensure cost-effectiveness and technical feasibility of the management of risk from contamination and the reuse of soil and debris.									X	
49.	4	50.	To improve social acceptance through identification and engagement of all stakeholders.	X									
61.	5	62.	To retain buildings and infrastructures on brownfield sites				X						
73.	6	74.	To reuse existing buildings and infrastructures, or components thereof, on brownfield sites.									X	
85.	7	86.	To recycle materials of existing buildings and infrastructures on brownfield sites									X	
97.	8	98.	To minimise energy demand and produce renewable energy on the site									X	
109.	9	110.	To minimise water demand and reduce waste water production									X	
121.	10	122.	To promote land use functions that match socio-economic demands and needs.				X						
133.	11	134.	To integrate the reuse of brownfield sites into regional land management				X						
145.	12	146.	To integrate the reuse of brownfield sites into urban development				X						
157.	13	158.	To achieve benefits for and prevent adverse impacts on the local neighbourhood.				X						
169.	14	170.	To generate and safeguard employment and economic development.			X							
181.	15	182.	To promote land use functions that suit the natural and man-made environment of the site and its neighbourhood.				X						
193.	16	194.	To save resources.									X	
205.	17	206.	To increase the possibility of the public traversing former brownfield sites.				X						
217.	18	218.	To provide adequate access.				X						
229.	19	230.	To achieve high urban design quality.									X	
241.	20	242.	To create and maintain flexibility and flexible urban design.									X	
253.	21	254.	To obtain better quality information.	X									
265.	22	266.	To improve information flow and use within the decision-making process.	X									
277.	23	278.	To deliver a fair discussion and conflict resolution process.	X									
289.	24	290.	To increase the legitimacy of the decision-making process.						X				
301.	25	302.	To improve the efficiency of the process in terms of duration and cost.	X									
313.	26	314.	To empower citizens, especially those representing non-organised interests.						X				
325.	27	326.	To delegate responsibility to lower decision levels and to stimulate a sense of ownership.	X									
337.	28	338.	To adopt an interdisciplinary project team approach.	X									
349.	29	350.	To facilitate efficient project delivery.	X									
361.	30	362.	To promote and manage stakeholder participation.	X									
373.	31	374.	To provide a framework for transparency in decisions, flow of information and improved communication structures.	X									
385.	32	386.	To protect human health and safety and the environment during site operations.	X									
397.	33	398.	To adopt an approach that integrates social, economic and environmental aspects.	X									

Figure 5.1 RESCUE sustainability objectives. *Source: RESCUE. 2005. Best Practice Guidance for Sustainable Brownfield Regeneration. Edwards, D., Bertram, C., Pahlen, G. & Nathanail, C.P. (editors). Land Quality Press: Nottingham.* Reprinted with Permission.

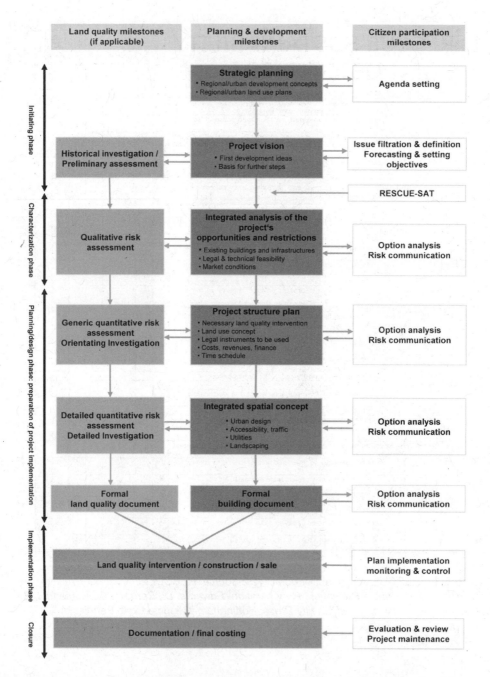

Figure 5.2 Decision chart (Radbod, Germany) illustrating the complex nature of brownfield regeneration projects. *Source: RESCUE. 2005. Best Practice Guidance for Sustainable Brownfield Regeneration. Edwards, D., Bertram, C., Pahlen, G. & Nathanail, C.P. (editors). Land Quality Press: Nottingham.* Reprinted with Permission.

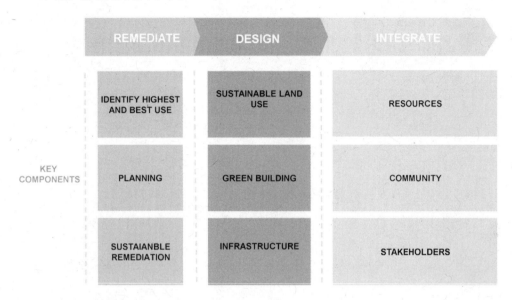

Figure 5.3   RE-DESIGN-IT framework.

CHAPTER
6

Land Planning

INTRODUCTION

Sustainable land planning and land use is the cornerstone of transforming brownfields into sustainable development projects. There are a number of factors that are driving land planning toward a more sustainable practice. Key elements of land planning and trends such as transit-oriented development will be explored in this chapter.

Today, we live in a time of dynamic technical progress and are also confronted with the consequences of that progress like the depletion of land resources that limit growth and other environmental consequences like global climate change. This experience is shared by nearly all countries in the world. There is a worldwide increase in the impoverishment of large groups of people because of increasing population and the related pressure on land resources.

With the shortage of available land resources, the search for effective land planning approaches have begun to take hold to sustainably manage land resources. Agenda 21, which was ratified by more than 170 nations at the Earth Summit in Rio de Janeiro in 1992, frequently refers to land use planning as playing a key role to natural resource management (http://www.un.org/esa/sustdev/ documents/agenda21/index.htm). Effective, sustainable land use planning provides a means to settle conflicts and conciliate interests which guarantee the sustainability of land resources. In this process, land use planning follows an integrated planning approach linking up various strategies by utilizing various technical disciplines.

As mentioned above, land planning encompasses various disciplines which seek to order and regulate the use of that land in an efficient and ethical way. That is, land planning is the scientific, aesthetic, and orderly disposition of land, resources, facilities, and services to ensure security of the economic and social efficiency, health and well-being of urban and rural communities (http://www.cip-icu.ca/web/la/en/pa/3FC2AFA9F72245C4B8D2E709990D58C3/te mplate.asp).

At its most basic level land planning involves zoning and transport infrastructure planning. In developed countries, land planning is an important part of social policy, ensuring that land is used efficiently for the benefit of the wider economy and population as well as to protect the environment.

Land planning encompasses the following disciplines:

- Architecture;
- Environmental planning;
- Landscape architecture;
- Regional Planning;
- Spatial planning;
- Sustainable Development;
- Transportation Planning;
- Urban design;
- Urban planning; and
- Urban renewal.

Architecture, urban design, urban planning, landscape architecture, and urban renewal address the selection of physical layout, scale of development, aesthetics, costs of alternatives, and selection of building materials and impact upon landscape and species.

Environmental planning addresses the implications of development and plans upon the environment. At the local level environmental planning may include the use of tools to model impacts (e.g., roadway noise, pollution, surface runoff, flood assessments) of development decisions.

To coordinate and exploit the various disciplines and expertise involved in land planning and to assist in the analysis and decision-making processes, land planners are increasingly utilizing information technology applications like geographical information systems (GIS) and special decision support systems.

Land planning over the past several decades has resulted in zoning ordinances that isolate employment locations, shopping and services, and housing locations from each other. Further, low density growth planning has created automobile access to increasing expanses of previously undeveloped land. These recent land planning practices have resulted in haphazard, inefficient, and unsustainable (sub)urban sprawl as demonstrated in Figure 6.1. The way we plan the physical layout, or land use, of our communities is fundamental to their sustainability.

The resultant complex problems of (sub)urban sprawl include traffic congestion and increasing commute times, air pollution, inefficient energy consumption, reliance on foreign oil, loss of open space and habitat, inequitable distribution of economic resources, and the loss of a sense of community. Community sustainability requires a transition from poorly-managed sprawl to land use planning practices that create and maintain efficient infrastructure, ensure close-knit neighborhoods and sense of community, and preserve natural systems.

LAND PLANNING STRATEGIES

As urban areas have grown and new land was used, land planning helped define the character of communities by regulating how land was used, determining where residential, commercial, industrial, and open space uses were to be located and developed.

However, as urban populations have swelled, with a majority of the population now living in cities (see Figures 6.2 and 6.3), it is no longer sustainable to simply consume new land for development to fuel the growth of the cities. Instead, attention has shifted to the use and re-use of previously developed land. This has led to the rise of and encouragement of redevelopment of brownfields as well as other land planning strategies.

Figure 6.1   An aerial photograph illustrating (sub)urban sprawl. (www.mbl.edu/news/features/images/aerial.jpg)

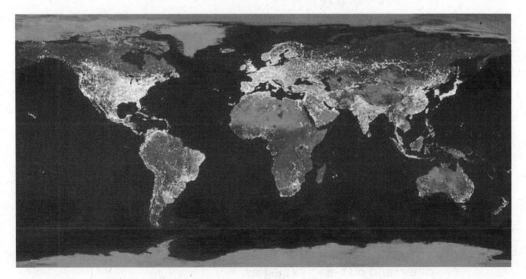

Figure 6.2   A map of the world at night. (http://apod.nasa.gov/apod/ap001127.html)

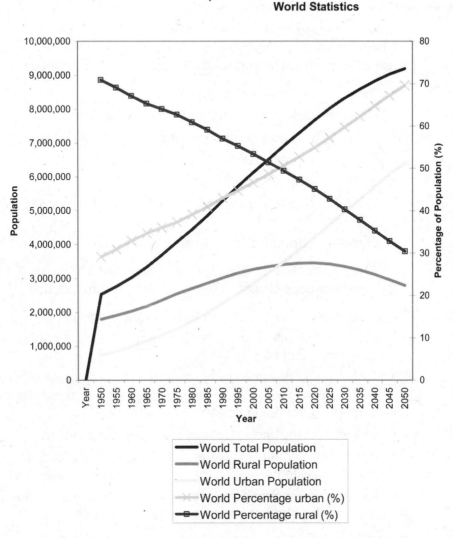

Figure 6.3 World population location statistics. *Source: World Urbanization Prospects: The 2007 Revision Population Database ©United Nations.* Reprinted with permission.

The consideration of a wide spectrum of factors including transportation, development density, energy efficiency, natural corridors and open space, and growth management are all required for sustainable land planning. The following essential strategies are components to address the complex land planning issues facing communities.

Transit-Oriented Development

Transit-oriented development (TOD) involves planning and design strategies for the development of mixed-use, walkable communities sited adjacent to transit access. Factors driving TOD include:

- Increasing traffic congestion;
- Growing "dislike" of suburban strip development;
- A desire for a quality urban lifestyle;
- Changes in public policy;
- Support for smart growth;
- Changing family structures (baby boomers becoming empty nesters); and
- Growing support for smart growth strategies.

Transit-oriented development, as illustrated in Figures 6.4 and 6.5, typically consists of the following design components:

- Walkability—the pedestrian as the highest priority
- Town center with train stations or other means of public transit as a prominent feature
- High density development with a mix of commercial office, retail, residential, and civic uses within close proximity of intermodal transportation nodes
- Transit systems including light rail, buses, trolleys, etc.
- Design that promotes use of bicycles, scooters, etc. as transportation systems
- Reduced or managed parking within 10-minute walk of development center and intermodal transportation nodes.

Transit-oriented development offers the following social, economic, and environmental benefits.

- Higher quality of life
- Better communities to live, work, and play
- Greater mobility
- Increased utilization of public transit resulting in reduced traffic congestion and driving
- Healthier lifestyles through the activity of walking to work and for errands

Figure 6.4   An element of TOD.

Figure 6.5   An illustration of transit-oriented development. *Image by Crandall Arambula.*
(http://www.ca-city.com/images/about_us/current_projects_lft.jpg)

- Reduced household spending on transportation
- Reduced stress within community members
- Increased foot traffic and customers for area businesses
- Reduced pollution
- Reduced cost of infrastructure.

Further specific benefits of transit include (Sustainability and Cities: Overcoming Automobile Dependence by Peter Newman and Jeffrey Kenworthy):

- "Transit investment has double the economic benefit to a city than does highway investment.
- Transit can enable a city to use market forces to increase densities near stations, where most services are located, thus creating more efficient subcenters and minimizing sprawl.
- Transit enables a city to be more corridor-oriented, making it easier to provide infrastructure.
- Transit enhances the overall economic efficiency of a city; denser cities with less car use and more transit use spend a lower proportion of their gross regional product or wealth on passenger transportation." (http://www.transitorienteddevelopment.org/tod.html)

Mixed-Use Strategies

Mixed-use development promotes the co-existence of many community services within close proximity to reduce automobile dependency. The ideals and realty of mixed use as a preferred urban form are still in their relative infancy as part of modern development. However, the ideals behind the push for this development form go back to the urban fabric of some of the oldest cities. The driving factors behind the attraction to the mixed-use strategy include current trends and attitudes toward a more urban lifestyle, changing demographics and psychographics, sustainable design trends and codification of the mixed use form in our development ordinances. Planning and design for successful mixed-use projects in today's environment is a complex matrix of skill sets and necessary collaborations between design professionals, developers, leasing agents, tenants, municipal planners, land use attorneys, local government, and communities.

Urban Growth Boundaries

Urban growth boundaries are a regulatory strategy for limiting urban sprawl by creating a geographical boundary for new development over a period of time, typically 20 years or more. Urban growth boundaries are used to control urbanization and minimize sprawl by designating the area inside the urban growth boundary for higher density urban development and the area outside the boundary for lower density rural development.

An urban growth boundary strategy typically consists of the following components:

- No sewer and water extensions to areas beyond the established boundary

- Reduced lot size and related zoning incentives for areas within the urban growth boundary and lot size disincentives for areas outside the boundary

- Subdivision regulations which require that developers pave town roads leading to new subdivisions

- Annual building permit limits for areas inside and outside the urban growth boundary, with project evaluation and rating criteria. Substantially more building permits would be allowed within the boundary than outside its boundaries. All development projects would be rated and permits awarded to projects best meeting community goals.

- Establishment of a community or regional fund to match state grants for land acquisition of farmlands or environmentally sensitive areas outside boundaries only.

Communities across the U.S. and globe utilize urban growth boundaries for land use planning. Some of the U.S. communities include Amherst, Massachusetts; Portland, Oregon; Goldendale, Washington; Cookeville, Tennessee; and many others. Even some U.S. states, like Oregon, have adopted legislation to allow for the creation of urban growth boundaries. Some other communities around the globe that utilize urban growth boundaries include Wyndham, Victoria, Australia; London, England; Copenhagen, Denmark; Vancouver, British Columbia, Canada; among many others.

Communities across the globe utilize urban growth boundaries for the following advantages:

- "Affirmation of a community's identity by ensuring that it doesn't merge with adjacent communities
- Promotion of urban and suburban revitalization
- Fiscal responsibility through more efficient use of public facilities
- Development of affordable housing
- Access to open space and greenways
- Stimulation of development that supports access to public transportation systems
- Stakeholder engagement by bringing together developers, government agencies, NGOs, and others to define areas of development and non-development; creating certainty about community growth.
- Encouragement of long-term strategic planning."
(http://greenbelt.org/downloads/about/ugb.pdf)

Figure 6.6 is an example of the urban growth boundary and zoning for Goldendale, WA.

Portland, Oregon's urban growth boundary has been in place since the 1970s (see Figure 6.7). The success of Portland's urban growth boundary strategy is evidenced by the preservation of forest and farmland at the region's edge; increased amount of housing planned inside the urban growth boundary from 129,000 homes to 300,000 homes; and revitalization of the downtown area. (http://greenbelt.org/downloads/about/ugb.pdf)

Infill Development

Infill development is a land planning strategy employed to promote greater development density and efficiency within existing urban boundaries. Successful infill development is effective at channeling economic growth into existing urbanized areas while conserving resources at the periphery. Infill development often includes:

- New development on vacant lots within urban areas;
- Redevelopment of underused buildings and sites; and
- The rehabilitation of historic buildings for new uses.

Over the last decade, the City of Chicago has promoted infill development and site reuse as exemplified in Millennium Park in Figure 6.8. Chicago's efforts have been characterized by industrial reuse

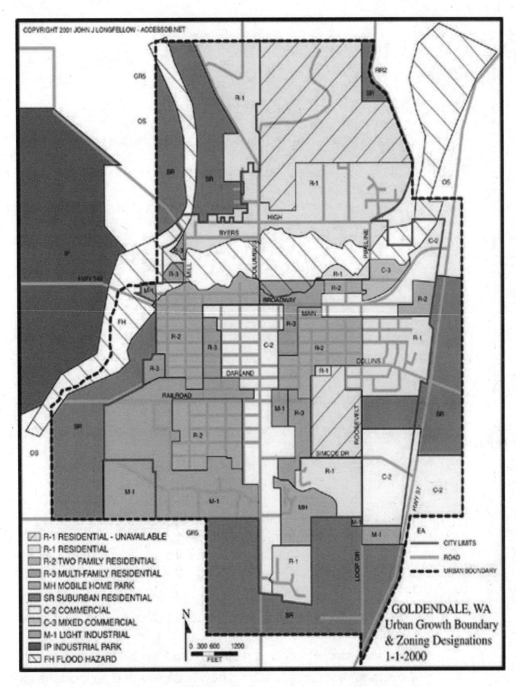

Figure 6.6 Goldendale, Washington urban growth boundary and zoning map. (www.ci.goldendale.wa.us/images/zoneMap.jpg)

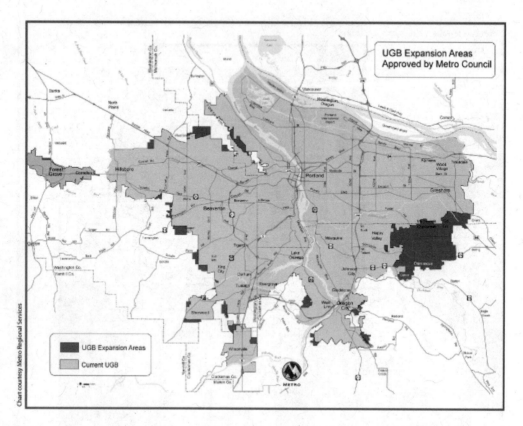

Figure 6.7 A map of Portland, Oregon's urban growth. *Courtesy of Metro-Regional Services, Portland.*

through the establishment of planned industrial districts and infill housing (http://www. nemw.org/infillch11.pdf).

In the mid-1990s, the City of Dallas, Texas adopted an infill development strategy to focus on the following strategies (http://www.nemw.org/infillch11.pdf):

- Preserve and build on the unique assets that make Dallas a desirable place to live
- Preserve, strengthen, and revitalize the foundation of community
- Leverage resources to attract new business and support the expansion of existing business

Figure 6.8   Downtown Chicago's Millennium Park.

- Transform the center city into a dynamic urban area, with a variety of business, cultural, entertainment, and living choices
- Strengthen Southern Dallas as an economically competitive and desirable place to work
- Protect and develop the Trinity River Corridor to become Dallas's new front yard—a nature park and a recreational and economic asset.

Figure 6.9 shows the Beaux Arts style Dallas Union Station that serves as the vibrant hub for Dallas Area Rapid Transit (DART), Trinity Railway Express, and Amtrak's Texas Eagle line.

Similar infill development strategies have been utilized at other urban areas throughout the U.S. with much success. Some of these cities include Denver, Colorado; Portland, Oregon; Manchester, Vermont; Orange County, California; and Christchurch, New Zealand, as Figures 6.10 and 6.11 illustrate.

Greenways

A greenways strategy is used to preserve open spaces and natural systems and provide recreational opportunities by connecting cities, suburbs, and rural areas through linear corridors such as parks and trails.

Many cities, counties, and states have their own citizen- or agency-led greenways programs to acquire land for greenways

Figure 6.9   Dallas Union Station. *Image by Dusty Garison.*
(http://www.trainweb.org/texasandpacific/collection/structures/Dallas_union_station.jpg)

Figure 6.10 An example of residential infill development in the Pearl District in Portland, OR.

and promote their use through programs and information as Figures 6.12 and 6.13 illustrate. A few examples are Florida's Office of Greenways and Trails, the grassroots groups Wachusett Greenways in central Massachusetts, Ozark Greenways in Missouri, and Mountains to Sound Greenway in Washington state, Mission Creek Bikeway and Greenbelt in San Francisco, and the city efforts by Asheville, North Carolina, Nashville Greenways, Bloomington, Indiana's Alternative Transportation and Greenways System Plan (PDF), and Indy Greenways in Indianapolis, Indiana. The Greenbelt Alliance works to protect the greenbelt of the San Francisco Bay Area. There are greenways programs in numerous other communities as well. (http://www.smartcommunities.ncat.org/landuse/greenway.shtml) (http://www.victoria.ca/cityhall/graphics/plnsrv_green_grnwys1.jpg)

Figure 6.11 Infill development map from Christchurch, NZ. *Courtesy of Christchurch City Council and DPZ Pacific.* (http://www.ccc.govt.nz/CentralCity/Programmes/InfillDevelopment.jpg)

The concept of greenways is not just an American land planning strategy. This concept is in use globally as illustrated in Figure 6.14. An example are the Prague-Vienna Greenways, a 250-mile long network of hiking and biking trails between Prague and Vienna. This trail system allows for users to walk or bike between his-

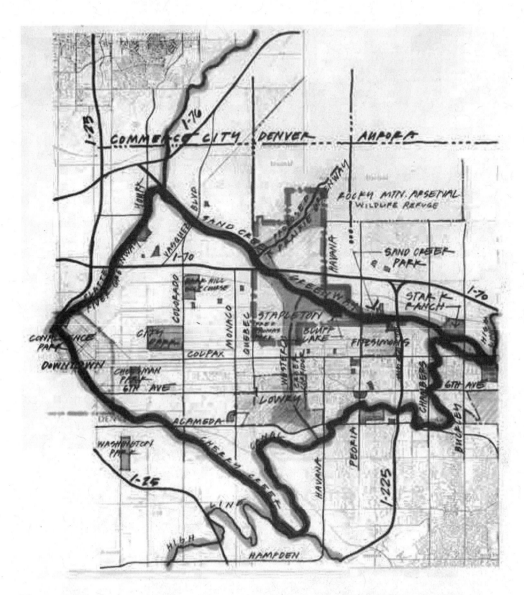

Figure 6.12 Illustration of the interconnectedness of greenways within the Denver metropolitan area. Here the Sand Creek Regional Greenway connects to the High Line Canal and the Platte River Greenway creating a 50-mile loop of off-street trails. (http://www.sandcreekgreenway.org/3map/3.2region/1region.html)

toric towns and villages, and explore historic cultural sites including castles, medieval churches and monasteries, etc. The routes stretch along the Vltava River Valley in Southern Bohemia and the

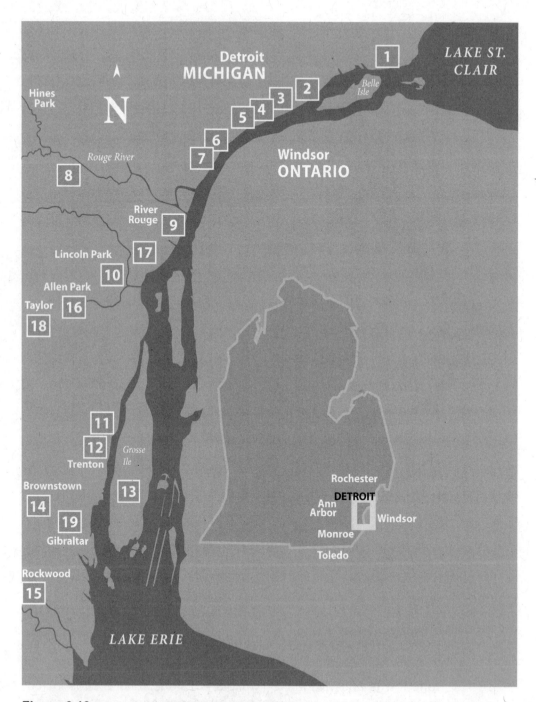

Figure 6.13   A map of the Detroit Greenways Project.
(http://www.miseagrant.umich.edu/greenways/images/Detroit-river-greenways.gif)

Figure 6.14   Experiencing the greenways.

Dyje River Valley in Southern Moravia (http://www.praguevienna-greenways.org/index.html).

The City of Victoria, British Columbia, Canada supports regional smart growth objectives to a pedestrian- and transit-friendly metropolitan core. The City of Victoria has established an urban policy of developing dense residential development within the downtown core of the city to enable residents to meet most of their daily needs without using a private, single occupancy automobile. In 2003, the City of Victoria Greenways Plan was adopted to provide a policy framework and a series of strategies to establish a greenway system throughout the city.

The City of Victoria has defined a greenway as a corridor of protected open space that is managed for conservation and/or recreation. The common characteristic of greenways is that they all go somewhere and follow natural land or water features (see Figure

Figure 6.15 Example of a greenway.

6.15). Further, the Victoria greenways are designed to provide a link between schools, parks, commercial areas, and recreation areas to encourage people to use some form of non-motorized transportation.

Transfer of Development Rights

The transfer of development rights (TDR) methodology is the exchange development rights among property developers to increase development density and protect open space and existing land uses. Specifically, TDR programs allow landowners to sever the development rights from a property and sell them to purchasers who want to increase the density of their developments. Further, local governments may buy development rights to control price, design details or restrict growth. TDR programs are used to preserve open space, agriculture, historic buildings or housing. (Transfer of Development Rights Programs; Jason Hanly-Forde, George Homsy, Katherine Lieberknecht, Remington Stone; http://government.cce.cornell.edu/doc/html/Transfer%20of%20Development%20Rights%20Programs.htm)

Transfer of development rights "is a device by which the development potential of a site is severed from its title and made available for transfer to another location (see Figure 6.16). The owner of a site within a transfer area retains property ownership, but not approval to develop. The owner of a site within a receiving area may purchase transferable development rights, allowing a receptor site to be developed at a greater density." (State of California, Office of Planning and Research, General Plan Guidelines, 1987).

TDR programs are advantageous as they:

• Compensate property owners to mitigate the impact of land use regulations

Figure 6.16 Transfer of development rights. (http://www.kingcounty.gov/environment/stewardship/sustainable-building/transfer-development-rights.aspx)

- Ease the implementation of zoning by providing a market for development rights
- Provide private funding to supplement public financing for the preservation of open space and historic sites
- Make development more predictable by creating clarity and consistency of zoning regulations through the purchase of development rights
- Utilize deed restrictions and conservation easements to permanently protect open space and historic sites.

King County, Washington TDR Program—King County's TDR program is a voluntary, incentive-based, market-driven approach to preserve land and relocate development growth away from rural areas and into urban areas. The program's basis on free-market principles and prices motivates landowner participation. Rural landowners receive economic returns through the sale of development rights to private developers that then build more compactly in designated areas. Since the year 2000, approximately 137,500 acres (215 square miles) of rural and resource lands have been protected from development by King County's TDR program (http://www.king-county.gov/environment/stewardship/sustainable-building/transfer-development-rights.aspx.)

Upper Blue Basin Transfer of Development Rights Program in Upper Blue Basin, Summit County, Colorado—The Upper Blue Basin TDR program allows private landowners in the rural backcountry of Summit County, Colorado to transfer the development potential of the property to areas more desirable for development purposes, primarily with the town of Breckenridge, Colorado. (http://www.sprawlaction.org/halloffame/IBluebasin.html)

A TDR Bank was established and is administered by Summit County to facilitate the implementation of the Upper Blue Basin TDR program. The TDR Bank creates a place for sellers and buyers to complete the transfer of development rights. Within the first 15 months of the program's operation, three transactions were secured resulting in the transfer of 15 development rights, preserving 300 acres of Upper Blue Basin backcountry. (http://www.sprawlaction.org/halloffame/IBluebasin.html)

Montgomery County, Maryland—Montgomery County, located adjacent to Washington, D.C., has a TDR program that involves buying easements, or restrictions, on thousands of acres of farmland to protect against the spread of residential housing developments. Additional acreage has been protected through a first-of-its-kind exchange program in which developers pay farmers to continue raising crops and animals. In return, the county gives developers permission to build extra housing units in otherwise controlled-growth areas. The Montgomery County program has resulted in the creation of a 93,252-acre agricultural preserve and 40,000 additional acres preserved through the exchange program. (http://www.washingtonpost.com/wp-srv/local/longterm/library/growth/part4/post4.htm)

Farmland Specific TDR Programs—(Table 6.1) presents a summary of local U.S. governments with TDR programs specifically for farmland.

Open Space Protection

Open space protection provides a means to protect a community's urban open space, farmland, wetlands, riparian lands, rangeland, forests and woodlands, and coastal lands. Open space (aka green space) is valuable to communities because it contributes to natural systems preservation, recreation, education, cultural heritage, and aesthetics (see Figure 6.17).

A few examples of U.S. governmental open space programs include:

- The Virginia Outdoors Foundation (http://www.virginiaoutdoorsfoundation.org) promotes preservation of open space through easements.
- The Peninsula Open Space Trust in San Francisco, California is dedicated to the permanent protection of the San Francisco Peninsula and Santa Cruz mountain range. (http://www.openspacetrust.org/about/index.html)
- The Georgia Community Greenspace Program (http://www1.gadnr.org/greenspace) provides for urban and rapidly developing counties and municipalities to permanently preserve 20 percent of their land and water utilizing state funds.

The Trust for Public Land (http://www.tpl.org) is a U.S. nonprofit working exclusively to protect land for human enjoyment and well-being. TPL helps conserve land for recreation and spiritual nourishment and to improve the health and quality of life in American communities.

Table 6.1   Summary of local U.S. governments with TDR programs specifically for farmland.
(http://www.farmlandinfo.org/documents/37001/TDR_04-2008.pdf) *continued on next page*

AMERICAN FARMLAND TRUST · FARMLAND INFORMATION CENTER

LOCAL GOVERNMENTS WITH TDR PROGRAMS FOR FARMLAND, 2008

Locality	Year of Inception	Rights Transferred	Agricultural Acres Protected	How Rights Are Used	Notes
California					
City of Livermore	2003	56 payments	$1,200,000	Increase residential density	Allows payments in lieu of transfers
Marin County	1981	11	660	Increase residential density	Multi-purpose program
Colorado					
Larimer County	1994	721	503	Increase residential density	Multi-purpose program
Mesa County	2003	10	50	Increase residential density	Multi-purpose program
Delaware					
Kent County	2004	157	157	Increase residential density Change permitted land use	Multi-purpose program
New Castle County	1998	93	300	Increase residential density	Multi-purpose program
Georgia					
City of Chattahoochee Hill Country	2003	21	21	Increase residential density Increase commercial square footage	Multi-purpose program Chattahoochee Hill Conservancy operates TDR bank
Idaho					
Payette County	1982	154	4,000	Permit development on substandard lots	Multi-purpose program
Maryland					
Calvert County	1978	UNK	13,260	Increase residential density	Multi-purpose program Purchases and retires rights
Caroline County	2006	136	1,500	Increase residential density	Multi-purpose program Maintains registry of interested buyers/sellers
Charles County	1992	1,110	3,330	Increase residential density	Multi-purpose program
Howard County	1993	NR	2,045	Increase residential density	Multi-purpose program Purchases and retires rights
Montgomery County	1987	9,630	51,489	Increase residential density	Operated bank but discontinued in 1990
Queen Anne's County	1987	UNK	8,032	Increase residential density Increase commercial square footage Increase impervious surface area	Multipurpose program Non-Contiguous Development activity included in county figures
St. Mary's County	1990	155	465	Increase residential density	
Massachusetts					
Town of Groton	1980	25	100	Increase residential density Increase rate of development	Multi-purpose program
Town of Hadley	2000	3 payments	$206,772	Increase commercial or industrial floor area Reduce parking requirements	Allows payments in lieu of transfers
Town of Plymouth	2004	13	118	Increase residential density	Multi-purpose program
Minnesota					
Blue Earth County	1996	150	6,000	Increase residential density	Multi-purpose program
Chisago County	2001	11	290	Increase residential density	Multi-purpose program
Rice County	2004	102	3,252	Increase residential density	Multi-purpose program
Nevada					
Churchill County	2006	200	688	Increase residential density	Multi-purpose program Operates TDR bank
Douglas County	1997	3,518	3,727	Increase residential density Increase commercial square footage	
New Jersey					
Chesterfield Twp., Burlington Co.	1998	652	2,231	Increase residential density Increase commercial square footage	Burlington County operates bank used by township
New Jersey Pinelands	1981	4,000	25,000	Increase residential density Permit development on substandard lots	Multi-purpose program Operates TDR bank Maintains registry of interested buyers/sellers

Table 6.1 Summary of local U.S. governments with TDR programs specifically for farmland. (http://www.farmlandinfo.org/documents/37001/TDR_04-2008.pdf) *continued from previous page*

LOCAL GOVERNMENTS WITH TDR PROGRAMS FOR FARMLAND, 2008

Locality	Year of Inception	Rights Transferred	Agricultural Acres Protected	How Rights Are Used	Notes
New York					
Central Pine Barrens	1995	48	48	Increase residential density Increase commercial or industrial density/intensity All permitted increases in density or intensity relate to, and are capped by, increases in sewage flow	Multi-purpose program Commission operates bank Maintains registry of interested buyers/sellers
Town of Perinton	1993	68	174	Increase residential density	Multi-purpose program Purchases and retires rights
Pennsylvania					
Honey Brook Twp., Chester Co.	2003	18	50	Increase residential density Increase non residential square footage Increase impervious surface area	
Manheim Twp., Lancaster Co.	1991	422	476	Increase residential density Increase commercial square footage Increase impervious surface area	Operates TDR bank Purchases and retires rights
Shrewsbury Twp., York Co.	1976	30	60	Increase residential density Allowance of certain non-residential uses	Operates TDR bank
South Middleton Twp., Cumberland Co.	1999	8	135	Increase residential density	Multi-purpose program
Warrington Twp., Bucks Co.	1985	187	UNK	Increase residential density Increase commercial square footage Increase impervious surface area	Multi-purpose program
Warwick Twp., Lancaster Co.	1993	447	897	Increase commercial and light industrial square footage	Operates TDR bank Partners with Lancaster Farmland Trust
West Vincent Twp., Chester Co.	1998	162	NR	Increase residential density Increase commercial square footage	Multi-purpose program
Vermont					
South Burlington	1992	414	497	Increase residential density	Operates TDR bank
Washington					
King County	2000	8	80	Increase residential density	Multi-purpose program Operates TDR bank
Snohomish County	2004	49	70	Increase residential density Increase commercial square footage	Operates TDR bank
Wisconsin					
Cottage Grove Twp., Dane Co.	2000	3	105	Increase residential density	
TOTALS		22,733	129,810		

Most of the programs listed in this table protect multiple resources including agricultural land. For the purposes of this table, we only included transfers from agricultural land and acres of agricultural land protected by each program.

Two programs included in this table—Livermore, Calif., and Hadley, Mass.—allow payments in lieu of transfers. For these programs, the figure in "Rights Transferred" column represents the number of payments received to date and the figure in the "Agricultural Acres Protected" column equals the funds received to date. These numbers are not included in the totals at the bottom.

UNK means the program manager did not know. NR indicates that the program manager did not respond.

Surveys were sent to programs identified by staff and profiled in publications and reports about TDR programs, including *Transfer of Development Rights in U.S. Communities:Evaluating Program Design, Implementation, and Outcomes* by Margaret Wells and Virginia McConnell and *Beyond Takings and Givings: Saving Natural Areas, Farmland, and Historic Landmarks with Transfer of Development Rights and Density Transfer Charges* by Rick Pruetz.

Figures for St. Mary's County, Md., are from the Wells/McConnell report. Figures for Queen Anne's County, Md., are from a presentation posted on the county's Department of Land Use, Growth Management and Environment Web site.

Figure 6.17 Example of open space.

The Center for Green Space Design (http://www.green-spacedesign.org) provides forums and educational tools to ensure smart design solutions for green space, advocating preservation of green space and implementation of quality patterns of growth.

Open Space Institute (http://www.osiny.org/site/PageServer) is a nonprofit land conservation organization that works to permanently protect from development landscapes of significant environmental, historical, and agricultural value in New York, Vermont, New Hampshire, and Maine.

Greenbelt Alliance (http://www.greenbelt.org) works to protect open space and promote livable communities in the San Francisco Bay area by protecting the region's greenbelt.

The Ecological Cities Project at the University of Massachusetts (http://www.umass.edu/ecologicalcities) promotes sharing of knowledge and experience regarding new approaches to urban greenspace creation and management.

The Land and Water Conservation Fund (http://www.nps.gov/lwcf) was created by Congress in 1964 to provide money to federal, state, and local governments to purchase land, water, and wetlands for the benefit of all Americans.

Urban Forestry

Urban forestry entails the planting and maintenance of trees within a city or community as a strategy for reducing both carbon emissions and energy expenditures for heating and cooling.

American Forests is the nation's oldest citizen-based conservation organization and a leader in the urban forestry movement. American Forests sponsors several programs including Urban Ecological Analysis, Global ReLeaf, and Cool Communities. The organization also convenes National Urban Forest Conferences and maintains the National Urban Forest Council, a national network of individuals active in the development of urban forests (http://www.americanforests.org).

The Heat Island Project of Lawrence Berkeley National Laboratory is an interdisciplinary team of researchers working together to find, analyze, and implement solutions to the summer warming trends occurring in urban areas, the so-called "heat island" effect. Part of the project examines the role of vegetation in cooling cities (http://eetd.lbl.gov/HeatIsland).

Friends of Trees is a nonprofit organization in Oregon whose mission is to promote the planting, care, and protection of urban trees. Another nonprofit group that helps plant and care for urban trees is Friends of the Urban Forest in San Francisco. Other community groups that help promote urban forests for their quality-of-life benefits are the Georgia Urban Forest Council and the Virginia Urban Forest Council.

New Jersey's "Cool Cities Initiative" is a tree planting campaign designed to address the urban heat island effect. The $5 million joint program of the Board of Public Utilities and the Department of Environmental Protection is planting 100,000 trees across the state (http://www.coolcities.us).

Figure 6.18   A tree lined residential street—an urban forest.
(http://www.treecanada.ca/programs/urbanforestry)

TreeVitalize seeks to develop a public private partnership, through a collaborative process, to address the loss of tree cover in the five-county Southeastern Pennsylvania region. Launched by the governor on Arbor Day 2004, the $8 million partnership plans to plant more than 20,000 shade trees and 2000 acres of forested riparian buffers (http://www.treevitalize.net).

As many as 78 percent of Canadians live in urban centers in which streets may resemble Figure 6.18 (http://www.treecanada. ca/site/?page=programs_urbanforestry&lang=en). The Canadian Urban Forest Network is an advocacy group for Canadian Urban Forestry whose stated mission is "to increase awareness of the urgent issues facing Canada's urban forests and to stimulate action to address those issues." The Canadian Urban Forest Network is seeking "a new deal for urban forestry" supported by the federal and provincial governments. The primary element of the "new deal" involves the Canadian government setting aside $24 million annually ($1 per person residing in urban centers of Canada) to assist in the ongoing support of Canada's urban forests (http://www.tcf-fca.ca/programs/urbanforestry/cufn/pages.php?lang=en&page=aboutus).

Several cities have developed urban forest programs that train citizen volunteers to help care for trees and educate the public about the value of urban forests. One example is the Citizen Forester Program in Washington, D.C. (http://www.caseytrees.org/education/citizen-forester/index.php)

Land Trusts

Land trusts are local, regional, or statewide nonprofit organizations directly involved in protecting important land resources for public and environmental benefit (see Figure 6.19). Land trusts may purchase land outright, accept donated land, or purchase or hold conservation easements or development rights on parcels owned by others. As one of the fastest growing conservation movements, over 1000 U.S. land trusts have protected millions of acres including farms, wetlands, wildlife habitat, urban gardens and parks, forests, ranches, watersheds, coastlines, river corridors, and trails (http://www.urbanstrategies.org/foreclosure/Community_Land_Trust/clt_flow.html).

The Trust for Public Land is a nonprofit land conservation organization working with government, business, and community groups to acquire and preserve open space to serve human needs, share knowledge of nonprofit land acquisition, and pioneer methods of land conservation and sustainable land use (http://www.tpl.org).

The Land Trust Alliance is a U.S. umbrella organization for land trusts that serves as an educator, coordinator, leader, and advisor to help land trusts save more threatened natural areas and open land (http://www.landtrustalliance.org). The Land Trust Alliance Web site provides a listing of land trusts by state that can help in locating organizations active in a particular area.

American Land Conservancy is a national, nonprofit organization that works in close partnership with communities, private landowners, local land trusts, public lands agencies, and elected officials to create effective conservation solutions for threatened land and water resources (http://www.alcnet.org).

One of the best-known and largest land trusts is The Nature Conservancy, which protects land worldwide (http://www.nature.org).

TRUSTEE   ◀

LAND TRUST
You as Beneficiary...or

(Better) Your Living Trust as Beneficiary

TRUSTEE   ◀

LAND TRUST
You as Beneficiary Assign to...

XYZ Corporation
Sell XYZ Properties

TRUSTEE   ◀

Corporate Trustee

LAND TRUST
You as Beneficiary Assign to...

Limited Partnership

Figure 6.19   A land trust diagram.

Community Land Trust Model

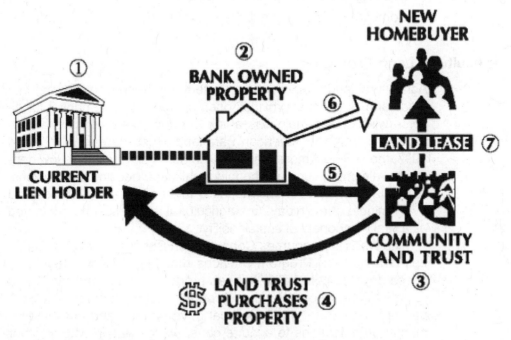

Figure 6.20 A community land trust model.
(http://www.urbanstrategies.org/foreclosure/Community_Land_Trust/clt_flow.html)

Many active and successful independent land trusts exist on the local, state and regional levels. A few examples include:

- Vermont Land Trust
- Maine Land Trust Network
- Gathering Waters Conservancy
- Marin Agricultural Land Trust

Another type of land trust, typically known as a community land trust, maintains ownership of land where affordable housing is constructed, as shown in Figure 6.20. Separating the cost of the land from the cost of the homes built on that land helps keep home prices affordable. The concept was introduced by the Institute for Community Economics, which provides information on how community land trusts work. Some examples of this type of land trust are:

- Rondo Community Land Trust in Minnesota
- Rockingham Area Community Land Trust in Vermont
- Homestead Community Land Trust in Seattle, Washington

Agricultural Land Protection

Agricultural land protection is a strategy for preserving the land that feeds and clothes us, provides open space, food and habitat for diverse wildlife, and maintains a link to our nation's agricultural heritage. According to American Farmland Trust, between 1992 and 1997, the U.S. lost more than 6 million acres of farmland to low density development such as shopping malls and housing subdivisions. They also say that America loses a further two acres of farmland every minute of everyday. Our agricultural lands must be preserved as a vital component of sustainability.

The Natural Resources Conservation Service Farm and Ranch Lands Protection Program provides matching funds to help purchase development rights to keep productive farm and ranchland in agricultural uses. Working through existing programs, USDA partners with state, tribal, or local governments and non-governmental organizations to acquire conservation easements or other interests in land from landowners.

The American Farmland Trust brings farmland loss and environmentally damaging farming practices to the attention of policymakers as well as the general public. The American Farmland Trust Web site provides an on-line Farmland Information Center, a newsletter, a searchable database of farmland protection statutes, and a series of publications and videos on farmland issues.

Glynwood Center helps communities work for economic development while preserving their natural and cultural heritages. Their Agricultural Initiative is involved in several projects designed to sustain small and mid-size farmers.

Rural Heritage Program of the National Trust for Historic Preservation is dedicated to the recognition and protection of rural historic and cultural resources. Through educational programs, publications, and technical assistance, the Rural Heritage Program supports the efforts of rural communities across the country to both preserve and live with their heritage.

Pennsylvania's Department of Agriculture Bureau of Farmland Preservation heads state efforts in agricultural land protection through purchase of agricultural easements. Pennsylvania leads the nation in the number of farms and acres of farmland protected.

The Marin Agricultural Land Trust is a land trust dedicated to preserving farmland in Marin County, California. MALT was the first land trust to focus exclusively on farmland preservation and is a model for efforts across the nation.

A number of states and counties have agricultural land preservation programs. For example: Kent County, Maryland; Delaware; New York; Michigan; California; and Lake County, Ohio. There are also numerous state and local organizations involved in regional farmland protection, such as: Colorado Cattlemen's Agricultural Land Trust; Skagitonians to Preserve Farmland; and Vermont Land Trust.

Solar Access Protection

Solar access protection utilizes legislation at the state and/or local level to enact measures to provide legal protection to property owners investing in solar energy systems through solar access ordinances.

On a broad scale, the State of Minnesota has required that all comprehensive development plans include protection for and development access to sunlight specifically for solar energy systems (http://www.metrocouncil.org/about/statutes.pdf).

A map highlighting U.S. states that currently have solar access laws at the state and/or local level may be viewed at http://www.statesadvancingsolar.org/policies/policy-and-regulations/solar-access-laws.

Green Building

INTRODUCTION

When one thinks of "greening brownfields" these days it is assumed the site will have green buildings. Certainly green buildings are an essential aspect of the greening of brownfield sites. However, "green development" is much more than green building as we will discuss in subsequent chapters.

Green building has taken off in the past several years due in large part to the creation and initiatives of organizations such as the U.S. Green Building Council (USGBC) and global green building organizations.

Green building is becoming more widespread both as a solution to specific building-related problems, and as a means of working toward a sustainable future. Green building programs are transforming the building development and construction markets. While some green building programs are sponsored by state or local government, others are administered by home building industry associations or by other nonprofit groups.

Green building is the practice of creating structures and using processes that are environmentally responsible and resource-efficient throughout a building's life-cycle from building siting, design, construction, operation, maintenance, renovation, and deconstruction. Green building expands and complements the classical building design concerns of economy, utility, durability, and comfort.

Green building is also known as a sustainable or high performance building that incorporates best practices, both financially and environmentally, in sustainable building. Green building is rooted in energy efficiency and resource conservation that inform decisions regarding design, construction, and operating practices. When implemented well, these decisions result in a win-win situation for a building's impact on the environment, occupants and community, and the owner's bottom line. Good green building practice includes considerations ranging from site selection, transportation access and infrastructure, material selection, energy systems, architectural design, landscaping, stormwater management, air quality, appliance and fixture selection, and more. Through an integrated approach to these strategies, green buildings will potentially:

- Reduce operational costs;
- Reduce resource consumption and contamination;
- Reduce indoor health related issues;
- Improve worker productivity;
- Increase net operating income;
- Increase resale value;
- Provide public recognition of demonstrated environmental commitment; and
- Demonstrate associated benefits to building occupants and the community.

There are currently numerous U.S. and global green building initiatives. It is interesting to note that in the U.S. not only is the USGBC working to build and transform the built environment, the USEPA is also on board and promoting green building practices. The involvement of the USEPA in promoting green building projects is very significant as it melds the historical USEPA focus on investigating and cleaning up contaminated properties (brownfields) with green development. The breaking down of these "silos" is encouraging and is part of an overall trend in integrated approaches to greening brownfield sites.

Let's examine some of the green building initiatives in both the public and private sectors.

U.S. LEGISLATION AND PUBLIC SECTOR GREEN BUILDING INITIATIVES[1]

The "explosive" growth in the green building industry can be illustrated by examining the U.S. green building market and public sector initiatives.

In the last several years there has been a ground swell of public sector interest in building and operating green. Throughout the United States, federal, state, county, and municipal governments have begun to incorporate energy and resource-efficient principles into public works programs. Together with a proactive federal government, they are starting to transform conventional practice in the building industry into an approach that reduces the environmental impact of construction-related activities while producing meaningful savings to the taxpayer. By championing green building practices in real estate development, the public sector is helping to set the standard for others to follow. Governmental organizations have also used their purchasing power to expand the markets for green building products, including clean and renewable energy technologies.

An important recent development has been the incorporation of the LEED Rating System principles and practices into the basic framework for state and municipal policies. In several cases, government agencies or appointed task forces have been working to supplement the national LEED standards with appropriate local laws and regulations. In this way, the U.S. Green Building Council's (USGBC) Leadership in Energy and Environmental Design (LEED) is rapidly becoming the U.S. standard for designing, constructing, and certifying sustainable buildings, as shown in Figure 7.1.

Under a 1999 Presidential Executive order and through programs fostered within many of its key agencies, the U.S. federal government committed to building green. The U.S. Armed Services (Army, Navy, and Air Force) and the General Services Administration (GSA), together with the Department of the Interior are using the LEED Green Building Rating System as the standard for renovating and constructing new facilities. At present,

Various LEED Initiatives

including legislation, executive orders, resolutions,
ordinances, policies, and initiatives are found in

43 states, including

190 localities (126 cities, 36 counties, and 28 towns),

33 state governments,

13 federal agencies or departments,

16 public school jurisdictions, and

39 institutions of higher education across the USA

Figure 7.1 SGBC LEED initiatives in the United States as of July 1, 2009.
Source: http://www.usgbc.org/DisplayPage.aspx?CMSPageID=1779.

several dozen public projects are certified or registered to achieve LEED certification.

With increasing frequency, mayors and governors are supporting the green building movement through executive orders or ordinances that commit their jurisdictions to building green. Some initiatives are voluntary, others compulsory, and a number offer building owners and developers incentives such as green building commercial and residential tax credits.

Table 7.1 provides examples of the LEED and green building legislation in U.S. states.

The U.S. Green Building Council (USGBC) keeps an updated file of LEED legislation in federal, state, and municipal governments. The information in Table 7.1 is from the February 2009 update located at: https://www.usgbc.org/ShowFile.aspx?DocumentID=691.

USEPA Green Buildings on Brownfields Initiative

USEPA's Green Buildings on Brownfields Initiative is designed to promote the use of green building techniques at brownfield properties in conjunction with assessment and cleanup. The USEPA is providing communities with technical assistance to facilitate the development of green buildings on their brownfields through pilot projects (http://www.epa.gov/brownfields/html-doc/greenbld.htm).

Eight communities have been granted Green Buildings on Brownfields pilot projects. USEPA's Green Buildings on Brown-

Table 7.1 Examples of LEED legislation by state.

STATE	POLICY TYPE	APPLICATION	RATING LEVEL	YEAR
Arizona	Executive order	State-funded buildings	Silver	2005
California	Executive order	New and renovated state-owned facilities	Silver	2004
Colorado	Executive order	State-funded buildings	Certification encouraged	2005
Maryland	Executive order	All captial projects greater than 5000 gsf	Certification encouraged	2001
	House and senate bill	State-funded capital projects	Silver	2005
New Jersey	Executive order	New schools	Certifiable**	2002
New Mexico	Executive order	Public buildings over 15,000 sf	Silver	2006
New York	Executive order	State projects	Certification encouraged	2001
Pennsylvania	Agency rule	Depts. of Environmental Protection and Conservation buildings	Silver**	NA
	House bill	Public school districts	Silver**	2005

Notes:
Possible LEED ratings: certified, silver, gold, and platinum
NA-Not available
* Based on costs
**USGBC certification not required
Source: U.S. Green Building Council

fields Initiative works with communities to incorporate environmental considerations into the planning, design, and implementation of their brownfields redevelopment projects. In general, the USEPA provides the following assistance:

- Technical
- Financial
- Planning
- Outreach
- Design

The eight Green Buildings on Brownfields Initiative pilot projects are presented below (http://www.epa.gov/brownfields/html-doc/greenbld.htm).

- *"Community Center Building, Springfield, Massachusetts.* With assistance from USEPA and local government, the New North Citizen's Council (NNCC), a nonprofit social service organization, plans to build a new, two-story, 25,000-square-foot building to replace its cramped and deteriorated offices. The project site is located on 1.2 acres, which includes current offices and several city-owned brownfield parcels. The pilot project aims to meet the criteria, and gain the certification, of the the U.S. Green Building Council's Leadership in Energy and Environmental Design (LEED) rating system and the Energy Star for Buildings program. The NNCC, which is developing the property in partnership with the City of Springfield, operates in two predominantly Hispanic neighborhoods with low per capita income and some of the highest asthma rates in the state.

- *The National Aquarium in Baltimore's Center for Aquatic Life and Conservation, Baltimore, Maryland.* To meet the needs of a rapidly growing collection of current and future exhibits, as well as expanding programs in research, husbandry, and conservation, the National Aquarium in Baltimore is building a new aquatic animal care center. A seven-acre brownfield site located near major highways and bus routes in Baltimore City is the planned location for the new facility. The aquarium will seek at least the silver or gold level of the U.S. Green Building Council's LEED rating system. A multi-use conservation education center, public access to trails, and boardwalks bordering wetlands are also proposed.

- *ReGenesis Medical Center, Spartanburg, South Carolina.* Re-Genesis, Inc., a community nonprofit corporation is purchasing a 33-acre brownfield for redevelopment as a health and wellness park. The property will include a community medical center providing integrated healthcare. Using green building technologies, the center will be designed to protect indoor air quality, reduce operation and maintenance costs, and protect the watershed of a nearby creek.

- *Marina District Redevelopment, Toledo, Ohio.* Dillin Corporation is in the process of redeveloping the 120-acre Marina District brownfield on the east bank of the Maumee River, directly across from downtown Toledo (see Figures 7.2 and 7.3). The redeveloped area will include residential, commercial, hospitality, recreational, and entertainment facilities. Dillin Corp. plans to incorporate into the site green building technologies such as energy efficient construction, a district-wide central heating system, natural landscaping, natural stormwater management, and a pedestrian-friendly boardwalk.

Figure 7.2 Conceptual drawing of Toledo's Marina District redevelopment project. (www.dillin.com/marina.html)

- *World Headquarters for Heifer International, Little Rock, Arkansas.* Heifer International, a nonprofit organization devoted to ending world hunger, is developing its new world headquarters and an education center on a 28-acre brownfield in a former industrial area in eastern downtown Little Rock (see Figures 7.4 and 7.5). Apparently, Heifer will seek a LEED Gold rating for the 100,000-square-foot building. Sustainable, environmentally sound features of the building will be highlighted in public education programs to illustrate environmentally responsible building practices. Programs will also demonstrate alternative energy, land and water conservation techniques, and efficient uses of natural resources.

- *Trailnet Inc., Trailhead Building, St. Louis, Missouri.* Trailnet, Inc., a nonprofit organization whose mission is to develop a system of greenways and trails in the greater St. Louis metropolitan area, is converting a former power plant building, located at the primary trailhead for the Confluence Greenway and Mississippi Riverfront trails, into an environmentally-friendly building to serve as office space for Trailnet. The building will also serve

Figure 7.3   Marina district redevelopment, Toledo, Ohio.

Figure 7.4 Heifer International's headquarters in Little Rock, Arkansas received the U.S. Green Building Council's highest rating, Platinum LEED™, for Leadership in Energy Efficiency and Design. *Photo courtesy of Heifer International/Tim Hursley.*

as a place for people to meet, dine, get directions and information on the Greenway and learn about the planet's third-largest watershed. Trailnet, working with local stakeholders and partners, aims to achieve a LEED Gold rating for the building (see Figure 7.6).

- *Community Culture and Commercial Center, Kauai, Hawaii.* This pilot project is located on the island of Kauai in Anahola on a property that has frequently been used for the disposal of items such as automobiles, tires, appliances, batteries, and other household items. Reuse plans for the state-owned property include a senior care living center, an elderly independent living facility, a charter school, retail, office space, and other support facilities. The state plans to design the buildings to be energy and resource-efficient, using local building materials wherever possible, and to attain LEED certification. Some of the

Figure 7.5   Another view of Heifer International's headquarters in Little Rock, Arkansas that received the U.S. Green Building Council's highest rating, Platinum LEED™, for Leadership in Energy Efficiency and Design. *Photo courtesy of Heifer International/Tim Hursley.*

sustainable design measures being considered include alternative energy generation, natural ventilation and day-lighting, water catchment, irrigation with gray water, use of recycled building materials, and the use of non-toxic finishes and materials.

- *Volcanic Legacy Discovery Center, Mt. Shasta, California*. The redevelopment plan for this 127-acre former lumber mill property includes 10 acres for the Volcanic Legacy Center. The center will be the centerpiece of a scenic byway from Crater Lake to Lassen Peak in California. The center will include a 20,000-square-foot building with a 200-plus seat auditorium, exhibit spaces with interactive and educational displays (including a section to demonstrate green building materials), a gift shop, other auxiliary spaces, and parking. In addition, sustainable landscape design

Figure 7.6   Trailhead Building in St. Louis, Missouri.
(http://www.confluencegreenway.org/locations/rt.php)

features may be incorporated into the reuse plan with a chaparral restoration area, vegetative filter area, wildlife pond, and stormwater retention basin.

U.S. COMMUNITY-BASED GREEN BUILDING PROGRAMS

Perhaps one of the most interesting aspects of the green building movement is the diversity of programs initiated at the local level. These local programs focus on unique sectors of the market and diverse community needs.

Examples of community-based (i.e., local government, NGO) green building programs within the U.S. are summarized below.

Several national initiatives are turning brownfields into green housing. One such initiative, Green Communities, is a five-year, $555 million effort led by Enterprise and the Natural Resources Defense Council (NRDC) to build more than 8,500 homes for low-income families that promote health; conserve energy and other natural resources; and provide access to schools, jobs, and services. Enterprise, NRDC, and other partners developed the Green Communities Criteria that provide a national standard for constructing green affordable housing, including the redevelopment of brownfields and the adaptive reuse of existing structures (http://www. greencommunitiesonline.org).

(http://www.urbanedge.org/green-housing.php)

In Massachusetts, the community development corporation Urban Edge relies on a mission- and market-driven approach to green housing on brownfields. Urban Edge developed green housing standards which it employs in all of its projects, almost all of which are built on brownfields. Egleston Crossing, Urban Edge's signature green building project, encompasses 64 affordable housing units and ground floor commercial space, and it was built on remediated brownfields (http://www.urbanedge.org).

Arlington County, Virginia, offers both a Green Choice Home program that promotes green residential construction and a Green Building Incentive Program that requires all site plan applications in the county to include a completed LEED scorecard (http://www. arlingtonva.us/Departments/EnvironmentalServices/EnvironmentalServicesMain.aspx).

(http://www.kitsaphba.com/bbk.html)

The "Build a Better Kitsap" program promotes environmentally friendly construction, energy-saving concepts, and use of natural resources in Kitsap County, Washington. The program establishes "code plus" standards to improve a building's performance and provide significant economic and environmental benefits to the owner and future generations. Features include point weighting for items, and a handbook with carefully linked content areas and local resources (http://www.kitsaphba.com/bbk.html).

The Master Builders Association of King and Snohomish Counties, Washington designed the Built Green program to provide Seattle, Washington area homeowners with comfortable, durable, environmentally friendly homes. The program's menu of actions that builders use to certify homes is the Built Green Checklist (http://www.builtgreen.net). The Built Green rating system includes points for building on a previously developed site, a brownfield site.

(www.builtgreen.org)

In the Denver, Colorado metropolitan region, Built Green Colorado was developed to highlight green builders. The program now applies statewide. The program is voluntary and serves as a guide for homes that meet certain green criteria. Builders and remodelers that participate in the program receive technical assistance, discounts on educational seminars, and other benefits (http://www.builtgreen.org).

(http://www.riversidebia.org)

Developed by the Building Industry Institute, the California Green Builder Program is a voluntary program that sets standards for improvements in energy efficiency, reduction in air emissions, onsite recycling, and reduction in water use (http://www.cagreen-builder.org).

(http://www.austinenergy.com/Energy%20Efficiency/Programs/Green%20Building/index.htm)

The City of Austin, Texas has developed a voluntary green building program. Austin Energy Green Building certifies green homes on a scale of one to four stars, with more stars being awarded to homes with more green features. The program also provides assistance to building professionals, such as architects, engineers, and builders, in exchange for offering and promoting green building practices (http://www.austinenergy.com/Energy%20Efficiency/Programs/Green%20Building/index.htm). The Austin Energy Green Building program has a requirement that the building project not be located on a previously undeveloped (greenfield) site (http://www.austinenergy.com/Energy%20Efficiency/Programs/Green%20Building/index.htm).

The City of Portland, Oregon has also developed a green building program as an integrated, conservation-based effort to promote resource-efficient building and sustainable site design practices throughout the City of Portland. Coordinating the expertise and resources of six City of Portland departments, the green building program sets aggressive goals and recommends a carefully selected set of strategies to leverage local expertise and develop cost-effective solutions for builders, developers, and building owners and users (http://www.portlandonline.com/osd/index.cfm?c=42133).

(http://www.earthcrafthouse.com)

Georgia's Greater Atlanta Home Builders Association created the EarthCraft House Program to provide training, design and construction technical assistance, marketing materials, and direct referrals to participating builders. Projects earn points by implementing a range of options for site planning, energy efficiency, waste reduction, water conservation and indoor air quality (http://www.earthcrafthouse.com).

(http://www.myfloridagreenbuilding.info/RatingSystems.html)

The Florida Green Building Coalition (FGBC) is a nonprofit Florida corporation dedicated to improving the built environment. Its mission is to create a statewide green building program that defines, promotes, and encourages sustainable efforts with environmental and economic benefits. The Florida Green Building Coalition has developed five certification programs (http://www.floridagreen-building.org/db):

- Green Home Standard
- Green Development Standard
- Green High Rise Standard
- Green Local Government Standard for Green Cities & Counties
- Green Commercial Buildings Standard.

Within FGBC's Green Development Standard, specific point opportunities exist for redeveloping a previously developed site with additional point opportunities for sites designated as a brownfield (http://www.floridagreenbuilding.org/db).

(http://www.gggc.state.pa.us/gggc/site/default.asp)

In the State of Pennsylvania, the governor's Green Government Council offers High-Performance Green Building Guidelines, as well as numerous online resources, reports, and a green building database to promote environmentally sustainable practices (http://www.gggc.state.pa.us/gggc/site/default.asp).

The City of San Jose, California created a Green Building program to encourage and facilitate new construction or remodeling of homes or workplaces into spaces that are healthier for people and the environment. The San Jose program promotes green building through the use of USGBC's LEED green building rating system (http://www.san-joseca.gov/esd/natural-energy-resources/greenbuilding.htm).

(http://www.ciwmb.ca.gov/GreenBuilding/Basics.htm)

The California Integrated Waste Management Board created a Green Building Design and Construction program to provide information for the sustainable construction of California state facilities including, a sustainable building tool kit for project managers and green building case studies (http://www.ciwmb.ca.gov/GreenBuilding).

(http://www.sbicouncil.org/index.cfm)

The Sustainable Buildings Industry Council collaborated with the National Association of Home Builders (NAHB) to create the Green Building Guidelines. Focused on residential, school, and commercial buildings, the sustainable design guidelines include illustrations, case studies, and checklists as a resource for builders and buyers interested in producing or purchasing resource-efficient homes.

(http://wi-ei.org/greenbuilt)

The Wisconsin Environmental Initiative has established the voluntary Green Built Home initiative that reviews and certifies homes that meet sustainable building and energy standards. This initiative is a collaboration between homebuilder associations, utilities, organizations that promote green building and energy efficiency, and the State of Wisconsin (http://wi ei.org/greenbuilt/index.php? category_id=3988).

In Boulder, Colorado the voluntary Green Points Building Program applies to both new construction and remodeling projects (less than 500 square feet). This program requires building permit applicants to earn "points" by selecting optional measures to receive a building permit. Homeowners and contractors are encouraged to include as many green options in their projects as possible (http://www.bouldercolorado.gov/index.php?option=com_content&task=view&id=208&Itemid=489).

(http://www.dnr.state.md.us/ed/index.html)

The Maryland Green Building Program works with county and municipal planners to evaluate and modify codes, ordinances, and policies to promote green building and development (http://www.dnr.state.md.us/ed/index.html). Maryland also offers a green building tax credit for commercial construction. One eligibility requirement for the tax credit is that the project must be located on a qualified brownfield site (http://business.marylandtaxes.com/taxinfo/taxcredit/greenbldg/default.asp).

NATIONAL GREEN BUILDING PROGRAMS

Let's take a look at the USGBC and global (national) programs.

The U.S. Green Building Council (USGBC) is the nation's foremost coalition of leaders from across the building industries who are working to promote buildings that are environmentally responsible, profitable, and healthy places to live and work. The USGBC launched the Leadership in Energy and Environmental Design (LEED) Rating System in 2000 as a third party verification and benchmarking tool for high-performance green buildings (see Table 7.2).

Each rating system is organized by prerequisites and credits. A project must meet all prerequisites and achieve a minimum number of credits to pursue or qualify for LEED certification. Depending on how many credits are achieved by a project, there are four levels of certification: certified, silver, gold, and platinum.

Under the current USGBC LEED program, one point is awarded for a green building on a brownfield site (http://www.usgbc.org/DisplayPage.aspx?CMSPageID=220). However, it is my understanding that the LEED scorecard may be revised to include a 110 point scorecard with more emphasis for developing on a brownfield site (http://www.usgbc.org/DisplayPage.aspx?CMSPageID=1849).

The LEED scorecard version 2.2 awards one point for developing a brownfield site and a total of 14 possible points for a sustainable site out of 69 possible points or 21 percent of the possible points. While all of the sustainable site points are not directly linked to brownfield redevelopment, they involve elements that are consistent with brownfield site redevelopment including development density, access to public transportation, and site selection.

The updated LEED 2009 scorecard keeps the credit for redeveloping a brownfield field site the same, but expands the scoring of related aspects such as access to public transportation and development density. The new scorecard increases the available sustainable sites points from 14 to 26, raising the value of a sustainable site to 24 percent of possible points. In fact, sustainable site points now represent 65 percent of the points required to reach LEED certified status for a new building, while in the past they only represented 54 percent. This demonstrates the rising importance of site selection as seen by the USGBC.

The Green Building Council of Australia (GBCA) was established in 2002 to develop a sustainable property industry in Aus-

Table 7.2   Summary of LEED registered and certified projects.
(http://www.usgbc.org/ShowFile.aspx?DocumentID=3340)

LEED	NEW CONSTRUC-TION	COMMERCIAL INTERIORS	EXISTING BUILDINGS	CORE & SHELL	NEIGHBOR-HOOD DEVELOP-MENT	SCHOOLS	RETAIL	TOTAL
Registered Projects	11,597	2047	2490	2488	225	713	189	19,524
Certified Projects	1600	479	200	157	13	4	36	2476

tralia and drive the adoption of green building practices through market-based solutions.

The goals of the GBCA are to drive the transition of the Australian property industry toward:

- Sustainability by promoting green building programs, technologies, design practices, and operations; and
- Integration of green building initiatives into mainstream design, construction, and operation of buildings.

To achieve these objectives, the Green Building Council of Australia launched the Green Star environmental rating system for buildings in 2003. Green Star rating tools help the property industry to reduce the environmental impact of buildings, improve occupant health and productivity, and achieve real cost savings, while showcasing innovation in sustainable building practices.

Green Star rating tools are currently available or in development for a variety of sectors, including commercial offices (design, construction and interior fit outs), retail centers, schools and universities, multi-unit residential dwellings, industrial facilities, and public buildings (http://www.gbca.org.au/green-star/rating-tools).

The Green Star Industrial Pilot program includes credit provisions specific to reclaimed contaminated land (i.e., brownfields). The purpose of this credit is to encourage and recognize developments that reclaim contaminated land that otherwise would not have been developed. Points are awarded where it is demonstrated that land prior to development is defined as contaminated, and where full adequate remedial steps have been taken by the developer to decontaminate or safely encapsulate the site prior to construction.

Taiwan Green Building Council

The "Green Building Promotion Program" of Taiwan provides a comprehensive mechanism of resources, research, guidance, training, and education to support the adoption of green building in Taiwan. The program emphasizes onsite ecological environment technology, construction waste reduction, building energy conservation, natural resource usage, indoor environmental quality control, and green building demonstrative projects (http://taiwangbc.org.tw/english/about.php).

New Zealand Green Building Council

The New Zealand Green Building Council is a broad-based industry organization formed to lead New Zealand's focus on green building. NZGBC is comprised of industry leaders who are committed to developing market-based solutions that help deliver efficient, healthier, innovative buildings for New Zealand (http://www.nzgbc.org.nz) (see Figure 7.7).

United Kingdom Green Building Council

The UK Green Building Council (UK-GBC) was organized to bring clarity, purpose, and coordination of sustainability strategy to the building sector.

Members of the UK-GBC include architects, engineers, investors, product manufacturers, cost consultants, energy suppliers, surveyors, developers, contractors, occupiers, NGOs, acedemic insitutions, and government agencies; stakeholders in the planning, designing, constructing, maintaining, and operating buildings (http://www.ukgbc.org/site/home).

BRE Environmental Assessment Method (BREEAM) is a voluntary measurement rating for green buildings that was established in the UK in 1990 as a tool to measure the sustainability of new non-domestic buildings in the UK.

Japan Sustainable Building Consortium

The Comprehensive Assessment System for Building Environmental Efficiency (CASBEE) formed is 2001 has compiled the following green building tools:

- New construction
- Existing building

Figure 7.7   One40 William is the first 5-Star Green Star building in Perth, Australia. *Image courtesy of HASSELL—Hassell.com.au.*

- Renovation
- Heat island
- Urban development
- Urban area and buildings
- Home.

Argentina Green Building Council

The Argentina Green Building Council utilizes the LEED rating systems.

Green Building Council Brasil

Green Building Council Brasil utilizes the LEED rating system, adapted to the Brazilian building sector.

Other established green building programs across the globe include the following:

- Emirates Green Building Council
- Canada Green Building Council
- Germany Sustainable Building Council.

EMERGING GREEN BUILDING PROGRAMS

Based upon the criteria of the World Green Building Council, the following councils have been accepted as emerging green building councils.

- Colombia Green Building Council
- Dutch Green Building Council—www.dgbc.nl
- Italy Green Building Council—www.gbcitalia.org
- Poland Green Building Council—www.plgbc.org
- Romania Green Building Council—www.rogbc.org
- Spain Green Building Council
- Vietnam Green Building Council—www.vsccan.org/vgbc.

The emergence of these global programs and the push by the USGBC are ensuring that green building is becoming mainstream. Green building practices are becoming part of public sector programs and I believe will ultimately become "building code."

Certainly in the U.S. LEED will continue to play a crucial role in the adoption of green building practices. However, as green building goes mainstream in the public and private sector.

REFERENCES

1. Includes excerpts from USGBC State and Local Government Toolkit, November 2002.

Valuing Financial Return or Beyond Asset Conversion

INTRODUCTION

In this chapter we will explore the approach to valuing the financial return on the greening of brownfield sites. In order to do this, we will first need to understand what the underlying motivations are for both private and public sector entities involved in the redevelopment process, and why these entities would elect to take this approach as opposed to a traditional development on a previously undeveloped parcel. We will discuss the concerns that these entities may have to this approach and identify solutions that have been developed to alleviate these concerns and make this approach attractive to the stakeholders involved.

The greening of a brownfield site is more than the remediation of a site and requires a more holistic approach to creating the full value from the site. Creating this full value requires that all stakeholders involved in any development (i.e., owner, developer, financier, insurer, investor, and the community or non governmental organizations [NGOs]) are striving to secure a financial return (or non- financial return) that is acceptable based on the capital that is placed at risk. Restoration of a brownfield site and redevelopment of the site utilizing green development concepts both introduce additional financial risks that must be overcome in order to make this type of development attractive to parties with capital at risk.

What makes redevelopment of a brownfield site challenging from a financial perspective is that there are additional risks that have been introduced as a result of prior use of the property, resulting in potential diminished property value. The property and buildings may be abandoned posing a hazard or may be obsolete, there may be significant soil, water, or air contamination, and/or the property may simply be underutilized with respect to the highest and best use of the property that would maximize its market value. This means that additional financial investment is likely in order to restore the value of the property back to a level that is equivalent with a previously undeveloped property before new development can take place.

In order to compensate for this additional risk associated with the remediation of the site, there must either be sufficient financial upside potential following redevelopment or some mechanism that reduces the level of risk.

If you speak to any brownfields developer they will tell you that the bottom line as to whether to develop a site or not (regardless of the remediation issues) is the location of the site (remember: "location, location, location"). Remediation and the greening of the site is just an additional aspect to manage in order to extract as much value as possible from the site. It is important to emphasize the concept of extracting full value, not just cleaning up the site or adding a green building.

Rebuilding the site with green development practices also introduces additional financial risks that are not typically encountered during traditional construction. The primary concern is often that the perceived higher upfront costs will not result in a sufficient increase in value to provide the desired financial return. There may also be concerns with regards to availability and durability of green products, knowledge with regards to design and implementation, or lack of builders familiar with the installation of green design elements. However, it appears that green construction has penetrated the market significantly enough that these are not as much of an issue as they were in the past. There is also the very real issue that many of the financial advantages of green development do not accrue equally to all stakeholders.

Many of these additional financial risks would be an insurmountable barrier to redevelopment without some other outside influence providing a catalyst that changes the financial viability of the project for those parties with capital at risk. This is where public sector and government involvement can be a critical component for overcoming many of these risks and improving the ROI for private stakeholders through both financial and non-financial support. By combining the resources of the public and private sectors, a brown to green redevelopment project can accrue the maximum benefit to the public sector while allowing private stakeholders to earn the desired financial return to make these redevelopment projects competitive with conventional development projects.

We are now seeing an alignment of public policy with private sector initiatives in green building. Historically, public policy has supported the redevelopment of brownfield sites and now that support is moving into green development. Green building is moving toward becoming municipal building code, and so the "option" of building green is looking less optional.

MOTIVATIONS, CONCERNS, AND SOLUTIONS

What types of entities might consider a green brownfields project, what are the motivations for completing the project, and what are the concerns and the potential solutions facing each entity to achieve an acceptable outcome?

The common image of a brownfield redevelopment project is an early 20th century heavy industry complex associated with the early years of the industrial revolution, located within the city core, which is radically transformed into a burgeoning mixed-the use development of middle-class housing within easy walking distance of new retail establishments and small company office space. While this is certainly a popular type of redevelopment scenario, there are many other types of brownfield sites, including former government facilities (including military bases), airports, shopping malls, schools, churches, commercial businesses, and landfills. The other popular image of a brownfield site is a property that is currently abandoned or has no economic activity occurring on the site. However, a viable brown to green project may also occur on a site where the economic activity is not living up to its full potential due either to obsolete facilities or obsolete uses. For example, a gasoline station in a bustling downtown area may not provide the maximum economic return for the site in a rapidly evolving economic climate. Although residential property could technically be considered as a brownfield site, it is not common practice for residential areas to undergo large scale redevelopment, so our discussion will only pertain to non-residential redevelopment.

The primary entities involved in the greening of a brownfield site will typically include the property owner or operator, a developer, and some type of financial or non-financial support from the public sector. Although this will often be the case, it should be noted that many different arrangements for ownership of the property may exist, including situations where the developer is also the owner, a government entity is the owner, or some combination of public-private ownership exists.

Public sector support may include one or more of the following stakeholders which may have an interest in the general welfare of the public or the environment:

- Federal government agencies (environmental, energy, and housing);
- State environmental agencies;
- State economic development and planning agencies;
- Local government agencies;
- Citizens and community groups;
- Local community development corporations; and/or
- Local, national, and global non-governmental organizations (NGOs).

Secondary private-sector participants fulfill a critical role in the successful implementation of a project but are usually third party vendors without an equity stake in the project. These participants may include one or more of the following:

- Commercial lenders
- Specialty brownfield development lenders
- Insurance companies
- Environmental consultants
- Remediation contractors

- Architects
- Engineers
- Green building professionals
- Legal counsel
- Real estate professionals.

Table 8.1 summarizes the typical motivations, concerns, and solutions for a typical brown to green redevelopment project which includes an owner/operator, developer, and government/public sector entities. Further discussion of the table elements are discussed below.

Owner/Operator Motivations

Owners and operators (businesses that operate on land that is leased from the owner) of non-residential property are primarily motivated to maximize the productivity and income generated from a piece of property. The decision to purchase a property or modify an existing property is predicated on the ability of the owner/operator to generate a desired ROI based on the capital invested. This ROI may be achieved based on the operating performance related to the use of the property or may be achieved based on the potential sale of the property and its income-generating potential.

As previously discussed, what makes redevelopment of a brownfield site potentially attractive to an owner/operator is that there are typically many favorable spatial and geographic factors associated with the site. Since brownfield sites by definition have been developed previously, they tend to occur within the existing infrastructure envelope of a city. The same features that attracted original development appeal to new development, including the presence of transportation infrastructure (road, rail, water), presence or access to water and sewer services, close proximity to markets for the products or services produced on the site, and access to labor markets. These are tangible features that can be difficult or expensive to replicate on a new site.

Another motivation particularly associated with inoperable or contaminated properties deals with an owner's/operator's ability to manage risk and liability associated with the property. Environmental risk can take the form of regulatory enforcement action and fines

Table 8.1   The motivations, concerns, and solutions in greening brownfield sites.

ENTITY	MOTIVATIONS	CONCERNS	SOLUTIONS
Owner/ Operator	• Maximize property productivity • Maximize property income • Maximize resale value • Maximize environmental liability • Risk management • Minimize operating costs • Access to transporation • Access to infrastructure • Attractive work environment (attract and retain workers, improve productivity) • Branding and reputational value creation • Address stakeholder and community issues related to sustainability	• Environmental liability • Worker health and safety • Unknown environmental conditions • Increased development costs • Demolition and cleanup costs • Permitting issues • Availability of financing and insurance • Making contamination issues public • Community resistance	• Retain control of liabilities and cleanup management • Voluntary cleanup programs • Insurance products • Specialized financing • Government-backed loans and tax incentives • Rebates and incentives for green building • Streamlined permitting • Reduced waste disposal costs • Community and stakeholder involvement
Developer	• Maximize project profitability • Reduce up front capital requirements • Reduce certain development costs (inputs, transport, waste) • Improve project marketability • Improve company image • Branding and reputational value creation	• Environmental liability • Worker health and safety • Unknown environmental conditions • Increased development costs • Demolition and cleanup costs • Length of time for cleanup • Permitting issues • Availability of financing and insurance • Marketability (contamination stigmas) • Community resistance	• Liability relief • Environmental site assessment • Innovative cleanup approaches • Insurance products • Specialized financing • Government-backed loans and grants • Knowledge dissemination • Rebates and incentives for green builiding • Streamlined permitting • Reduced waste disposal costs • Community and stakeholder involvement
Public Sector/ Government	• Create jobs (green jobs?) • Environmental justice issues • Increase tax revenue • Decrease crime and vandalism • Eliminate blighted properties • Spur revitalization • Increase surrounding property values • Prevent spread of pollutants • Reduce health hazards • Reduce waste to landfills • Capitalize on existing infrastructure • Land and ecological preservation (reduce sprawl) • Preserve natural capital	• Ensuring project completion • Default on loans • Allocation of tax revenue for competing interests • Community resistance	• Ensure established, reputable, and financially secure participants • Surety bonds • Evaluation of benefits to the community • Community involvement

associated with contamination, exposure of site personnel to chemicals and contaminants that might cause adverse health issues, and the potential for migration of contaminants off-site which could adversely affect human health or the environment leading to third-party legal issues. For inoperable properties, risk can take the form of vandalism, theft, or the potential for injury to trespassers.

The motivations for implementing green building practices for the owner/operator are tied to maximizing income by reducing operating expenses (reduced lifecycle costs) and improving worker productivity. The owner/operator of a property that is responsible for the operating costs of the property has a keen interest in reducing those costs in order to maximize profits. Incorporation of green building components and practices provides an opportunity to save costs both during construction and during the operational life of the building for materials use, energy use, water use, wastewater generation, stormwater runoff, and waste sent to landfills. Many of these practices can be implemented without additional costs, such as orienting a building to maximize solar heating in cooler climates or minimize solar heating in warmer climates. Additionally, green buildings provide an attractive and healthier work environment for employees. Including green building features can make it easier to attract and retain employees, and can increase productivity due to greater comfort and decreased illness.

Another potential motivator for owner/operators to green a brownfield site is improving brand and reputational value to its stakeholders, including employees, customers, investors, the local community, and NGOs. Companies that are seen as good stewards of the environment often enhance their image with stakeholders, and this perception can decrease costs by making it easier to recruit and retain employees and enhance a "license to operate." By proactively addressing issues related to sustainability, companies can reduce potential exposure to adverse situations such as shareholder resolutions and community activism. Green branding can often build a competitive advantage in the marketplace resulting in an increase in revenue due to customers who prefer to do business with companies that are seen as environmentally responsible.

Owner/Operator Concerns

Owner/operators have many concerns when it comes to greening a brownfield site that they might not encounter on a traditional development project. These concerns are primarily centered on the additional costs potentially associated with site rehabilitation and green redevelopment, and whether these costs can be recovered by the value added by using this approach. How the value is actually calculated is also critical as it is difficult (but not impossible) to determine brand and reputational value.

Environmental issues are of concern as they relate to environmental liability, worker health and safety, the presence of unknown environmental conditions, and the cost of demolition and remediation. Former owner/operators who are involved in redevelopment of a site are likely liable with regard to environmental contamination resulting from past operations; however, a new purchaser/user of the property will have concerns with regard to the potential transfer of environmental liability for past site uses. The owner/operator may be concerned about the potential for third-party legal claims as a result of off-site migration of contamination, worker health and safety during cleanup operations, and long-term liability associated with any off-site disposal of wastes. There is also a concern about the potential for unknown environmental conditions that are discovered during cleanup or redevelopment activities that could delay the project or add costs. Demolition and cleanup costs can be a significant percentage of redevelopment costs depending on the nature and extent of contamination, and cleanup activities often require the use of specialty contractors, equipment, and technologies that can add a premium to site development costs and extend the duration of redevelopment activities. Many owner/operators are also concerned about addressing environmental contamination issues publically as they fear that this could lead to bad publicity or litigation.

Green building "concerns" (really perceived concerns and biases) relate primarily to the perceived additional costs for incorporating green building concepts. Owner/operators may be concerned that green construction may require the use of specialty consultants and contractors and that the additional coordination between the various contractors and suppliers will add to the overall project costs. However, this concern is fading as green building goes mainstream. There are also concerns that certain building materials (e.g., additional insulation) and building practices (e.g., segregating waste material for recycling) may cost more than traditional construction materials and methodologies.

Other concerns may include costs and time for additional efforts to research and obtain building permits, especially in jurisdictions that are unfamiliar with green building features (e.g., living roofs). Requests for changes to zoning could also create barriers to redevelopment in the case of land use changes, although this is also changing as green building becomes building code (the Chicago green roof program is an example of building code changes in response to leadership from Mayor Daley).

The ability to obtain financing through traditional commercial lenders and obtaining insurance for non-traditional building features could also be a concern for owner/operators. Many banks and insurance companies may be reluctant to get involved in brown to green projects due to fear of transfer of environmental liability or unfamiliarity with green building products. Finally, community resistance is always a concern when large-scale changes to an existing site are proposed. However, if stakeholders are proactively engaged on a truly sustainable development project, they can be allies.

Owner/Operator Solutions

Solutions that are available to owner/operators either partially or completely alleviate many concerns that these entities have with regard to brown to green projects. The solutions are focused primarily on reducing the financial risks and uncertainties, and some of the direct costs potentially associated with site rehabilitation and green redevelopment. These solutions can help restore a favorable ROI with regard to these projects.

When addressing environmental liability, owner/operators have several options that can help control the liability and keep costs under control. Both new and historical owner/operators of a site should consider retaining control and management of the cleanup and redevelopment activities to help ensure that activities are carried out as planned and that costs are managed. For new owner/operators within the U.S., the federal Small Business Liability Relief and Brownfields Revitalization Act of 2001 provides liability protection for parties who purchase a property but who were not responsible for environmental contamination. There are also various government programs at the federal, state, and local levels that can provide both technical assistance as well as financial assistance for cleanup activities in the form of loan guarantees, tax breaks, and grants. State voluntary cleanup programs provide a regulatory platform for cleanup of facilities with a great deal of flexibility for how cleanup goals are established and cleanup is achieved. The insurance industry has also responded to brownfield cleanup issues with insurance products that help limit liability, provide a cost cap for environmental cleanup, and provide coverage for unexpected or undiscovered environmental conditions. Owner/operators should consider hiring qualified environmental professionals who can help identify environmental issues and propose cost-effective and innovative options that can reduce costs and cleanup duration. By proactively cleaning up a site and managing communication with the public, owner/operators can take actions that reduce the risk for contamination to spread off site and can alleviate much of the dissemination of misinformation that tends to alienate and anger the public.

Although certain costs for implementing green building concepts do add to project costs (capital or first costs), there are many other features of a green project that can offset those costs. It is important that the project proceed as a coordinated effort between all contractors involved in the project, preferably under the direction of a green building consultant. Coordination of efforts from the beginning of the project tends to minimize additional costs associated with this approach.

It is critical that when remediation efforts are "scoped" that the end use (or several scenarios) are kept in mind. This end use will inform remedial alternatives and the sequencing of these remedial actions.

Reuse of a brownfield site has many advantages during the redevelopment phase. The site is likely to have much of the infrastructure that will be required for the new development which will reduce site development costs. Many of the materials from demolition of obsolete structures can be reused which saves on the purchase, transport, and waste disposal costs. Additional costs associated with making the building more energy efficient (e.g., additional insulation) are often offset by enabling the installation of smaller and less expensive cooling and heating equipment. There are also various government programs (in the U.S. at the federal, state, and local levels) that can provide both technical assistance as well as financial assistance for the addition of green building features in the form of rebates and tax credits.

By engaging the public in the redevelopment process, and by potentially including local or state government agencies in the process, issues associated with zoning and permitting can be streamlined since the goals of the owner/operator and the community will tend to be in better alignment. Commercial banks and insurance companies have become more willing to lend and insure brown to green projects due to legislation that protects them from past environmental liability, and as these institutions begin to link sustainable practices to decreased risk. Additionally, there has been an increase in the availability of specialized financing relating directly to the implementation of brown to green projects in recent years which include private investment firms that can provide direct financing or are willing to take an equity position in the project. Finally, there are various government programs that can provide loan guarantees that aid in obtaining financing.

Developer Motivations

The primary motivation for a developer to engage in a brown to green project is the ability to maximize value (top line, bottom line, and brand value). The ROI for greening a brownfield site must be competitive with a traditional project, or even more advantageous from a financial perspective, to overcome the potential perceived (or real) additional risks. Developers may be attracted to greening a brownfield site because of reduced up front capital requirements. If a developer is interested in purchasing a property to be redeveloped, the property can often be obtained at a significant discount due to the real or perceived diminished value of the property. In other cases, there are landowners who are willing to provide an equity share in the property to the developer so that the developer may participate in the potential financial upside of the completed redevelopment.

Developers may also be attracted to the reduction in certain costs associated with the development. Much of the infrastructure for the site may already be in place (e.g., water, sewer, etc.) and recycling of onsite structures can reduce the cost of purchasing inputs, transportation, and waste disposal. Because these sites are usually located within urban areas, there is often improved access to suppliers, labor, and transportation.

As the concept of sustainability has begun to take hold in the marketplace, many developers have found that sustainable development projects appeal to more and more people. This is resulting in improved marketability of these types of projects in many markets. Many developers also see this as an opportunity to brand themselves as sustainable builders to capitalize on this movement, and to provide a new offering and potential edge over their competition. There is also an opportunity here for developers to enhance their image in the communities in which they work as being a good steward of the environment and performing projects that benefit the community by helping to preserve open space by reducing sprawl and conserving resources.

Consumer surveys indicate that even in the depths of a recession (2008, 2009, and ?) consumers and businesses view sustainability as a key issue and in some instances would pay a premium. One only needs to look at the Bank of America building in New York City developed by the Durst Organization (www.durst.org) and the success they have had with a high occupancy rate despite the "soft" commercial real estate market.

Surveys

One interesting point to note is that developers do not have any incentive when it comes to reduction of operational costs or lifecycle costs associated with the development, unless the developer is going to take a longer term equity position in the development. However, this is not a common practice as developers typically prefer not to tie up capital for periods beyond the completion and sale of the development. It is therefore important that the developer is aware of the motivations of the owner/operator and is willing and able to accomplish the goals needed by the owner to achieve his or her long-term ROI.

Developer Concerns

Developers have many of the same concerns as the owner/operator when it comes to performing a green brownfields project that they might not encounter on a traditional development project, particularly if the developer purchases the property or takes an equity stake in the project. These concerns are primarily centered on potential liabilities and additional costs associated with using this approach.

Because a developer typically will not (and cannot) tie up capital for significant periods of time, the developer is particularly concerned with the length of time that the project will take which can be affected by difficulties in getting approvals and the duration of any cleanup activity. Because a developer is not the historical owner/operator of the property, they do not have first-hand knowledge of prior operating practices and the potential for the extent or magnitude of any potential contamination issues. The developer will also be concerned as to whether the property has a stigma that might influence the marketability of the project, and may be more concerned than an owner/operator when it comes to current economic and market conditions due to his or her shorter investment horizon.

Developer Solutions

Many of the solutions that apply to owner/operators are also applicable to developers. Because developers were not responsible for prior environmental issues, they are generally protected from liability in the U.S. under the federal Small Business Liability Relief and Brownfields Revitalization Act of 2001. Developers who have purchased the parcel to be redeveloped should consider retaining control and management of the cleanup and redevelopment activities to help ensure that activities are carried out as planned and that costs are managed. A developer who is not familiar with the past operational practices at a facility will want to ensure that sufficient due diligence is performed by qualified environmental professionals to quantify the extent and magnitude of environmental issues and determine reasonably accurate estimates on the cost and duration of any cleanup activities. In the U.S., developers should also look to enroll the site in a state voluntary cleanup program which can provide a flexible mechanism for cleanup with valuable oversight and result in a no further action letter being issued upon completion of cleanup activities. This letter can be a powerful tool for helping to overcome skepticism about the environmental condition of the property when it comes to various stakeholders such as lenders, insurers, future owners, and the local community. Furthermore, developers can be instrumental in reducing development costs by working with environmental professionals to come up with innovative cleanup solutions. For example, paved areas or building footprints can be strategically located to effectively cap certain types of contamination thereby eliminating or reducing the cost of treatment and/or disposal of contaminated soils.

Developers should also be aware that there are typically additional sources of grants, loan guarantees, and incentives available for cleanup of a site for an owner who was not responsible for the contamination. Insurance products that provide a cost cap for environmental cleanup and provide coverage for unexpected or undiscovered environmental conditions should also be explored. As with owner/operators, conducting a community outreach program can pay significant dividends by encouraging community support and reducing potential delays associated with cleanup and redevelopment activities. Oftentimes these types of projects can be showcase properties that spur additional redevelopment providing potential future business opportunities for developers.

As with owner/operators, developers should look for opportunities to reduce construction costs during green building activities. It is critical to involve green building professionals and coordinate efforts early on in a project to minimize the potential for higher construction costs. Developers can save costs by utilizing existing infrastructure, reducing the amount of materials brought to the site and waste generation, reducing the amount of transportation of materials, and work with the green building professional to reduce equipment costs. For example, a more energy efficient shell for the building or a green roof results in reduced heating and cooling needs for the building, therefore smaller equipment can be specified. Developers should also seek out the various government programs that can provide both technical assistance as well as financial assistance for the addition of green building features in the form of rebates and tax credits.

Similar to owner/operators, developers should implement a communications program with the public and local government which can facilitate and reduce the time needed to streamline zoning and permitting issues. Developers should seek out those commercial banks and insurance companies that have become more willing to lend and insure brown to green projects, or should pursue specialized financing through private investment firms who can provide direct financing or are willing to take an equity position in the project. Government programs should also be explored for loan guarantees and financing if obtaining private lending is difficult or costly.

Government/Public Sector Motivations

Unlike the private sector participants in a green brownfields redevelopment project, the government/public sector does not have profit as the primary motive for participating in these types of projects. The primary motive for the government/public sector is in the ability of these entities to provide for the welfare and greater good of society. By making these projects more attractive to the private sector through financial and non-financial means, the government/public sector is able to accomplish its mission. In this case, ROI is not a measure of financial accomplishment but becomes a measure of the value provided to society as a function of the investment involved.

The motives for the government/public sector center on stimulating economic development while providing protection for human health and the environment. As properties in an area move through the typical real estate lifecycle of growth, stability, and eventually decline, underutilized or abandoned properties begin to appear which lead to a downward spiral in economic activity within the affected area. This decline is marked by loss of jobs, decreasing local tax revenue, decreasing property values, and increasing incidents of crime. Private investment is busy chasing opportunities in areas of growth and therefore investment in these declining communities continues to decrease. In order to reverse the continuing decline in economic activity, there must be a catalyst to make investment in these areas more attractive and competitive with other investment opportunities. By providing the proper incentives to the private sector, the public sector is able to stimulate the economy by spurring revitalization of an area in decline. Not only does this provide short term employment during the cleanup and redevelopment phase, but it provides sustainable longer-term jobs that benefit the surrounding community. It might be said that a brownfield redevelopment is contagious because revitalized properties tend to attract more investment and redevelopment activity as an area in decline undergoes revitalization and enters a new growth phase. This new growth phase increases the tax base, decreases crime and vandalism, and increases the value of surrounding properties. A once blighted area can now support a thriving community with growing employment, healthier tax revenues so that government can provide for the general welfare of its citizens, and improved financial security and quality of life for the community.

During the decline phase of an area, environmental quality also suffers as uncontrolled contaminants in soil, air, and water may migrate from source areas into the surrounding community. This may result in increased risks to human health and risks to ecological systems, which extol a higher cost on society. By providing a finan-cial incentive for cleanup and redevelopment, the public sector is able to accomplish the goal of addressing potential health issues and continuing damage to the environment without resorting to lengthy and expensive enforcement actions which do nothing to improve economic activity.

Other benefits to the environment and resource protection are multifold. By reutilizing property that has already been developed, the public sector is able to reduce urban sprawl and ecological de-terioration caused by development in new areas. In-fill projects can capitalize on existing infrastructure which means that building of new facilities (e.g., wastewater treatment) and transportation infra-structure can be avoided. The geographic location of previously developed sites near transportation networks, suppliers, workers, and customers also reduces fuel used for transportation and makes it more efficient for public transport systems to service the facility. Green building benefits may include reducing waste sent to landfills, reducing water use, reducing or eliminating stormwater runoff, reducing energy use, recycling of materials, and improved indoor air quality. Green buildings are healthier places to work and typically lead to reduced worker illness and increased worker pro-ductivity. All of these have trickle down benefits to society, includ-ing reducing the need for additional infrastructure (e.g., water supply, wastewater treatment, stormwater facilities, power genera-tion), reduced mining, extraction and transportation of raw materi-als associated with the site and necessary infrastructure, and reduction of pollutants to soil, water, and air, including greenhouse gas emissions which adversely impact the climate.

Government/Public Sector Concerns

The primary concern of the greening of a brownfield site from the perspective of the public sector is whether the investment will ac-tually produce the desired results. There are many competing needs for public sector funding that must be weighed against the benefit that they provide and there may be community resistance that must be overcome. For example, members in a community may feel that resources might be better allocated to improving the education system or building a new school as opposed to provid-ing investment in a private business that stands to have direct financial gains on the backs of taxpayers. The public sector must also consider the risk of default by the recipients of public funding, whether that results in loans that can't be repaid, a project that doesn't reach completion, or a project that fails at achieving the desired objectives.

This appears to be shifting in favor of promoting and investing in greening brownfield sites as a way to increase revenue and develop new jobs (green jobs), in particular in the U.S. (refer to the previously mentioned Sustainable South Bronx initiative as an example).

Government/Public Sector Solutions

As with owner/operators and developers, it is critical that public entities engage in communications with the communities that will be impacted by brown to green redevelopments. This can be done by implementing a public outreach program, soliciting input, and responding to concerns. This is also an opportunity to demonstrate to the community the shared benefits and to point out how this will improve the quality of their lives. With regards to default or projects achieving the goals set out by the public sector, it is important that public sector entities ensure that the owner/operator, developer, and other private entities involved in the project are reputable and financially secure. Surety bonds can be required from the owner/operator or developer as an insurance policy against nonperformance.

APPROACH TO VALUING THE FINANCIAL RETURN OF THE GREENING OF BROWNFIELD SITES

The approach to valuing the financial return of greening a brownfield site is similar in many respects to how the return is valued on a traditional development project. However, a brown to green project includes additional aspects of valuation that are significant but not typical of traditional projects. Actual calculation of ROI is beyond the scope of this discussion as the approach to analyzing financial returns can vary substantially based on the needs of the investors. Also, we will not discuss the specifics of how to calculate the value of greening a brownfield site (for the same reason as it varies according to the developer) versus traditional projects based on the similar components, but will instead focus this discus-

sion on those attributes that are unique to green brownfields redevelopment which need to be taken into consideration to properly
evaluate these projects from a financial perspective.

Direct measurable costs associated with most if not all of the aspects of the property cleanup phase of the development can be calculated and factored into standard ROI evaluations. The total cost
of cleanup should be calculated net of any financial incentives (e.g.,
grants, reimbursement programs) that are obtained for this phase of
the project. The desired ROI for the project can then be added to the
net cost for rehabilitation to determine what the required market
value of the property will need to be in order to achieve the desired
ROI (a real estate appraisal of the property can be performed using
a hypothetical condition to get an estimate of what the property
value might be following the rehabilitation phase of the project). The
component costs that may need to be considered during the rehabilitation phase of the development include the following:

- Environmental site assessment (ESA)—Although traditional development properties will require a phase 1 ESA (in the U.S.)
  which typically includes a physical inspection of the property, review of historical and surrounding land use information, and a
  review of government and environmental databases, a brownfield property will typically require a substantially more thorough
  ESA. The Phase II ESA will consist of additional inquiry into the
  historical uses of the property and operational details of past
  businesses including the use, storage, handling, and disposal of
  potentially hazardous materials. The investigation will also include the collection and analysis of samples to determine the
  presence of contaminants in soil, air, water, or building materials (e.g., asbestos, lead paint) in order to evaluate the magnitude and spatial extent of any potentially hazardous materials.
  These services can be performed by a qualified environmental
  professional and the cost will vary based on the extent of the investigation that is required.

- Voluntary cleanup program (VCUP)—If cleanup actions will be
  required, there are costs associated with enrolling the site in a
  state VCUP program and ongoing costs for regulatory oversight
  in the U.S. The VCUP program will provide review of the
  cleanup work performed and will provide a "no further action"
  letter upon successful cleanup of the property. The applicant
  should also inquire about any financial reimbursement pro-

grams that may exist which can pay for some or most of the costs of cleanup for certain types of releases (e.g., release from a petroleum storage tank or release of dry cleaning chemicals).

- Demolition—Costs for demolition of any structures will need to be factored in, if applicable. This may include costs for removal and disposal of any hazardous materials, costs for segregating materials for reuse or recycling, and costs for off-site disposal. These costs can be estimated by a qualified contractor.

- Cleanup activities—Once the nature and extent of any contamination is defined, cleanup activities may include costs for feasibility studies, pilot studies, preparation of plans, design for remedial activities, removal activities, ex-situ transport and disposal of materials, in-situ treatment of soil and/or water, and potential long-term groundwater monitoring. These costs can be estimated by a qualified environmental professional or remediation contractor.

- Insurance—Calculate the cost of any insurance obtained for capping cleanup costs, providing liability protection, or providing coverage for unknown environmental conditions, if applicable. These costs can be estimated by an insurance provider.

- Financing—Calculate the portion of any financing costs associated with this phase of the redevelopment. These costs can be estimated by a lender or investor.

- Community outreach—This activity may consist of conducting public meetings, sending out mailings, performing door-to-door surveys, posting information in media publications, setting up of a document repository for public review, and setting up of a Web site for dissemination and collection of information. These costs can be estimated by internal personnel or by a consultant.

A new purchaser of a property must take these costs into consideration when determining the purchase price of a property. Because many of these costs may be poorly defined prior to purchase of a property, caution should be used when negotiating a purchase price so that a sufficient discount is included to account for the cost needed to bring the property up to the value of a comparable unimpaired property. Sufficient due diligence should be performed prior to any purchase and the buyer may want to include a contingency clause or escrow of funds from the sale for additional protection against financial surprises associated with environmental conditions.

Calculating the ROI for the green building phase of a project is more complicated because most of the costs are tangible and easily calculated while many of the benefits are intangible and difficult to calculate. The fact that many of the costs are encountered during the development phase, but much of the cost savings are realized during the operational lifecycle of the building makes it critical that lifecycle costs are used to determine the ROI. Atypical costs associated with green redevelopment of a property can include many of the following, which may cost more than traditional construction, although not always with careful planning (LEED-NC Version 2.2 Reference Guide, U.S. Green Building Council):

- Additional architectural and engineering
- Pollution prevention during construction
- Bicycle storage and changing rooms
- Restoration of habitat and open space
- Vegetated roofs
- Pervious paving
- Stormwater capture and reuse system
- Landscaping to prevent heat buildup
- Water efficient fixtures
- Composting toilet systems
- Gray water capture and reuse systems
- On-site wastewater treatment systems
- Building commissioning
- Energy efficient HVAC systems
- Energy efficient lighting
- Other energy efficient equipment
- Improved insulation and sealing of buildings
- Energy efficient windows
- Installation of renewable energy
- Recycling of waste materials
- Utilizing recycled or renewable materials
- Utilizing certified wood
- Enhanced ventilation systems
- Air quality monitoring systems
- Indoor air quality management

- Use of low VOC materials
- Additional natural lighting
- Controllability of systems.

While savings resulting from things such as reduced energy use, water use, and waste reduction can be easily measured and calculated over the life of the development, it is much more difficult to calculate the savings of things such as missed work due to illness and increased worker productivity. Research in the area of trying to quantify these benefits has resulted in a wide range of values and correlations; however, it will take additional analysis before more accurate information is available. Suffice it to say that the benefit is greater than zero, and in some cases may be the single largest portion of the ROI.

Examples of Financial Returns

I would like to focus briefly on the financial benefits of building green as this is typically the primary focus of any "greening" strategy. There is much discussion regarding the cost of green building and, unfortunately, less about the value (both tangible and intangible). This is changing as the financial benefits of building green become better documented. While the following studies are focused on U.S. green building projects the principles apply to green building projects anywhere.

The key study in the U.S. regarding green building costs and value was conducted for the California Sustainable Building Task Force (http://www.cap-e.com/ewebeditpro/items/O59F3259.pdf and http://www.acigmvc.com/complete%20speakers%20presentation.pdf) and evaluated the additional capital costs and lifecycle costs for 33 green buildings.

This is currently the most comprehensive study performed in the U.S. and worth spending some time reading as it has become the basis for much discussion regarding "first costs" and value of green buildings. In summary, the results of the study were that an upfront premium of approximately 2 percent ($3–$5 per square foot) resulted in lifecycle cost savings of more than 10 times the initial investment assuming a conservative 20-year life for the buildings. The average energy reduction from the 33 buildings was 30 percent. This alone provides savings sufficient to pay back the initial 2 percent premium in less than nine years. The same study found that, over a 20-year period, the overall net savings for a green building is between \$48.87–\$67.31 per square foot, depending on the LEED rating of the building. Therefore, an initial investment of only 2 percent of the first costs results in savings worth more than 10 times the added premium.

Even if the benefits of worker illness and increased productivity are excluded, there is still a four times return on the original investment. This is a far lower first cost investment premium than is commonly perceived and clearly the benefits are worth considering.

Some of the summary graphics best illustrate the results of the study (http://www.cap-e.com/ewebeditpro/items/O59F3481.pdf).

First, an overview of typical versus green building segment. Figure 8.1 highlights the spectrum of the building market and that green building (focusing on LEED) is still an emerging market opportunity. Practitioners are essentially the risk takers.

The bottom line on the results of the study are summarized in Table 8.2.

The benefits of building green are illustrated as a percentage of total costs in Figure 8.2. It is critical to note that productivity improvements are the largest portion of the benefits in building green and this is usually the segment most ignored. However, this value

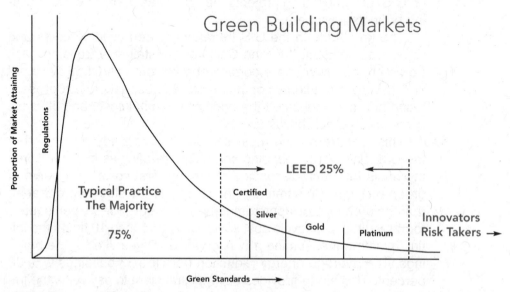

Figure 8.1   Overview of the green building market.

Table 8.2 Financial benefits of green buildings. *Source: "The Costs and Financial Benefits of Green Buildings—A report to California's Sustainable Building Task Force, October 2003."*

Financial Benefits of Green Buildings (per ft^2)	
Category	
Energy Value	$5.79
Emissions Value	$1.18
Water Value	$0.51
Waste Value (construction only - one year)	$0.03
Commissioning 07M Value	$8.47
Productivity and Health Value (Certified and Silver)	$36.89
Productivity and Health Value (Gold and Platinum)	$55.33
Less Green Cost Premium	($4.00)
Total 20-year NPV (Certified and Silver)	**$48.87**
Total 20-year NPV (Gold and Platinum)	**$67.31**

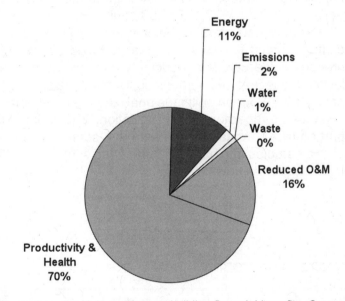

Figure 8.2 Percentage breakdown of green building financial benefits. *Source: "The Costs and Financial Benefits of Green Buildings—A Report to California's Sustainable Building Task Force, October 2003."*

is increasing not only as worker productivity and retention become increasingly important and valued by developers (especially, multi-national corporations), but by employees looking for employment opportunities (increasingly important to "millennials").

Of course there is currently (though this will likely not always be the case) a premium for green building depending on the "level" of effort put into integrating green building attributes, illustrated in Figure 8.3. If one looks at the LEED system it is clear that the higher the certification level, the greater the investment.

Increased productivity as a major benefit of green buildings should not be undervalued as people spend approximately 90 percent of their time indoors, and the concentration of pollutants indoors is typically higher than outdoors, sometimes by as much as 10 or even 100 times (U.S. Environmental Protection Agency, "Indoor Air Quality," January 6, 2003. Available at: http://www.epa.gov/iaq and Judith Heerwagen, "Sustainable Design Can Be an Asset to the Bottom Line—expanded Internet edition," Environmental Design & Construction, Posted 07/15/02. Available at: http://www.edcmag.com/CDA/ArticleInformation/features/BNP__Features__Item/0,4120,80724,00.html).

The numerous studies indicate that the benefits include significantly reduced illness symptoms, reduced absenteeism, and increases in perceived productivity over workers in a group that lacked these features. In addition, two studies of over 11,000 workers in 107 European buildings analyzed the health effect of worker-controlled temperature and ventilation. The report relies in large part on recent meta-studies that have screened tens or hundreds of other studies and have evaluated and synthesized their findings.

Figure 8.3   Average green premium versus level of green certification for offices and schools.

The reasons for improved productivity are straightforward:

- Much lower source emissions through avoiding locating air intakes next to outlets, such as parking garages, and avoiding recirculation), and better building material source controls.
- Significantly better lighting quality, including improved daylight harvesting and use of shading, greater occupancy control over light levels, and less glare.
- Generally improved thermal comfort and better ventilation—especially in buildings that use underfloor air for space conditioning.

Finally, two studies, conducted by the Heschong Mahone Group in Sacramento, California (http://www.thespaceplace.net/articles/hobstetter200703.htm) demonstrated that incorporation of natural light had positive results on a building's occupants. One study analyzed test score results for over 21,000 students from three elementary school districts in California, Colorado, and Washington state. Results from the Capistrano Unified School District in Orange County, California, indicated that in one year students with the most daylighting in their classrooms progressed 20 percent faster on math tests and 26 percent faster on reading tests. Another study examined sales levels in 108 nearly identical retail stores, of which some incorporated skylights and others did not; results showed 40 percent higher sales in daylight retail environments.

The bottom line is that extracting full value from a brownfield site means taking into consideration brand value, increased worker productivity, reduced operating costs, etc. It is difficult to quantify this full value but the smart developers factor in these benefits when they are greening brownfield sites.

Case Studies

INTRODUCTION

There is no shortage of excellent case studies to illustrate how brownfield sites are undergoing a green transformation. The challenge was to find the "iconic" projects that best illustrate the movement to green brownfield sites—those sites that are not just "built green" but that are transformational.

Wherever possible, information regarding the following aspects of the projects was explored.

- Why this brownfield site?
- Why build green?
- What was the process for evaluating the viability for brownfield redevelopment and deciding to build green?
- What was the community outreach and engagement process?
- What was the economic evaluation process for the site and did it include any value or premium for being green?
- How were the development partners "convinced" to develop green?

Instead of moving through key case studies organized by geography, the examples are presented in two major categories: "evolutionary green" and "revolutionary green."

What is evolutionary green? For our purposes it is essentially more of a transitional improvement in greening brownfield sites—not "cutting edge" or "bleeding edge" in vision and approach. Instead, improvements in sustainability performance and best practices characterize this group of case studies.

On the other hand there are several truly transformative projects that move brownfield redevelopment well beyond just green building or sustainable land use. These projects are "revolutionary green" and provide a vision beyond step improvements. These project developers and stakeholders are exploring what is possible with regard to "near zero" footprint and "social performance."

I view revolutionary green projects as those that follow the One Planet Living (OPL) framework and also those that leverage brownfield transformation for social good, including the creation of "green jobs." In my mind the most sustainable projects are not just focused on environmental footprint, they must create a social benefit and align with an extended group of stakeholders.

Let's first set the stage with a review of evolutionary green projects.

EVOLUTIONARY GREEN

Atlantic Station

I would like to start with the Atlantic Station project in Atlanta as it remains a very high visibility green brownfields project. The project not only effectively reused a brownfield site but has moved to a green development project (www.tndtownpaper.com/Volume8/evolution_of_brownfield.htm).

As of 2006 the project was the largest brownfields redevelopment project in the U.S. at over 138 acres in midtown Atlanta (see Figure 9.1.) The site was formerly the location of the Atlantic Steel Company active from the 1920s through the 1980s.

The initial challenge at the site was focused on remediation. The steel mill was demolished starting in July 1999 and quickly moved into the site remediation phase. Contaminated soils and materials consisted of approximately 9000 dump truck loads (approximately

165,000 tons), a groundwater interception system was installed, and stormwater detention facilities were constructed (including a one-acre pond in the center of the residential development). Ultimately the entire site was covered with clean fill materials.

After the remediation, the project got interesting and started to move in the direction of becoming a green brownfields project. As part of the negotiations with USEPA, a charrette was held with the USEPA, the Georgia Department of Transportation (GDOT), and the developers. The charrette was run by Duany Plater-Zyberk & Co. (DPZ) and was successful in reshaping the project into an innovative mixed-use project satisfying the needs of the stakeholders.

Figure 9.1   A view of Atlantic Station in downtown Atlanta, Georgia. *Photo courtesy of Atlantic Station.*

This moved the project along and the Atlanta City Council is-sued a $75 million tax allocation bond to pay for the first phase of infrastructure development. This was followed by the construction of a bridge facilitating access to the project site and the project was underway.

One of the key issues was access to the development from Atlanta. The northern boundary of the site is the main line for Nor-folk Southern Railroad and the south side is cut off from the neigh-borhood of Home Park. The eastern boundary is bounded by I-75/I-85 with the 17th Street Bridge as the solution to access.

As with most green building projects, there is a person or organ-ization that is driving the vision. Brian Leary had the vision to see something different for Atlantic Station. According to Leary, a vice president at Atlantic Station, in a 2006 interview ("Evolution of a Brownfield" in 2006, www.tndtownpaper.com), "We started focus-ing on creating places for people. We traveled all over looking for examples of good urbanism, trying to unravel the DNA of good places—why they work. We started referring to what we are doing as 're-urbanism.' Leary essentially moved the site from a classic "big-box" project to a "macro mixed-use."

And Leary continues, "The vision is to create an urban place for people, one that allows for evolution over time" and "more than 3000 people will live here by the end of summer 2006. We're a long way from being done or even halfway done though. When we're finished, more than 10,000 people will live here, and more than 12 million will visit each year."

The project specifications are (from the Atlantic Station Press Kit, www.atlanticstation.com):

- 138-acre project developed by AIG Global Real Estate Invest-ment Corp. and Atlanta-based Jacoby Development, Inc. (JDI).
- The community is projected to include 15 million square feet of retail, office, residential, and hotel space as well as 11 acres of public parks.
- The Atlantic Station community represents $2 billion in new construction and is divided into three areas:
  - The District
  - The Commons
  - The Village
- These three areas combined will provide:

- Six million square feet of class A office space
- 3,000–5,000 residential units (for sale and for rent)
- 2 million square feet of retail and entertainment space, including restaurants and movie theatres
- 1000 hotel rooms
- 11 acres of public parks

• The District will:

- Serve as the center of the community and will exemplify the live-work-play theme of the Atlantic Station neighborhood. A true mixed-use environment, The District will provide retail, entertainment, small- and large-scale office space, residential lofts above retail, townhomes, and other living options.

- The District will ultimately include:

  - More than 2 million square feet of open-air retail and entertainment
  - Six mixed-use retail buildings with entertainment, shops, and restaurants
  - 6 million square feet of office space in mid- to high-rise office buildings
  - An additional 150,000 square feet of office space atop retail locations
  - 156 loft apartments above retail
  - 100 for-sale townhomes and single-family detached homes

• The Commons will:

- Be the residential hub of Atlantic Station with a minimum of 1150 units surrounding a two-acre lake. The Commons also includes "artifacts" of the former steel mill.

• The Village will:

- Consist of 400–600 apartments/lofts adjacent to retail, occupied by residents with mixed incomes.
- Include a 366,000-square-foot IKEA store.

As with these types of projects, financing was from a variety of sources:

• Tax allocation district (TAD) bonds issued under the Georgia Redevelopment Act. Established in 1985, the Georgia Redevelopment Act was developed by city, county, and state leaders mirroring laws in other states that provide incentives for the development of land which is in need of economic incentives for redevelopment. The act provides incremental property tax, generated from future development, over the site's historic tax base to pay debt service on the bonds. Proceeds from the bonds can be used to pay for qualified development costs, as outlined in the law.

- Loans were provided by Wachovia Bank, Regions Bank of Atlanta, and Compass.
- Bank of Birmingham and Detroit-based Comerica Bank.

The general timeline for development of the project (also from the Atlantic Station Media Kit) is provided below:

- 1901—Atlantic Steel Co. founded as Atlanta Steel Hoop Co.
- 1915—Company became known as Atlantic Steel
- 1920s–1980s—Active steel mill
- 1974—With 2000 employees producing three-quarters of a million tons of steel annually, the company was Georgia Power's largest customer—consuming enough power to supply a city of 60,000 people
- 1979—Atlantic Steel Company acquired by Ivaco, Inc.
- Spring 1997—Property contracted to Jacoby Development, Inc. (JDI)
- April 1998—Atlanta Mayor Bill Campbell approves redevelopment zoning
- October 1998—EPA accepts Project XL TCM proposal
- December 1998—Atlantic Steel Mill closes
- June 1999—17th Street/Atlantic Station, LLC TCM is approved ARC ITIP—for improvement to regional air quality
- July 1999—Atlantic Steel begins deconstruction of steel mill
- September 1999—The Atlantic Steel redevelopment becomes the nation's first EPA Project XL for transportation and land use
- Fall 1999—Remediation begins
- December 1999—JDI closes on the Atlantic Steel property and forms partnership with AIG Global Real Estate Investment Corporation

- September 2000—The Atlantic Steel redevelopment project is named "Atlantic Station"
- March 2001—Beazer Homes joins the Atlantic Station redevelopment to build homes on the property, Lane Co. chosen to develop multifamily housing in Atlantic Station redevelopment
- October 2001—Atlanta City Council issues $75 million tax allocation bonds to pay for the first phase of infrastructure development and Georgia Department Of Transportation (GDOT) approves 17th Street Bridge design and construction
- November 2001—GDOT issues request for bids for construction of 17th Street Bridge
- December 11, 2001—EPD grants NFA (no further action) to Atlantic Station development
- December 14, 2001—17th Street Bridge contract let

Figure 9.2 A view of Atlantic Station in downtown Atlanta, Georgia. *Photo courtesy of Atlantic Station.*

- 2002–2003—Atlantic Station community begins announcing lease agreements; California Pizza Kitchen and United Artists first to sign on. Other tenants currently with signed lease agreements include Dillard's, IKEA, Publix, and Rosa Mexicano.
- October 2003—First residents move to the Atlantic Station neighborhood
- January 2004—17th Street Bridge construction is completed
- April 2004—17th Street opens into the Atlantic Station redevelopment and the first office tower (171 17th Street Building) opens with construction beginning on the IKEA store
- June 2004—Retail construction begins in The District
- August 2004—Groundbreaking for boutique hotel/residential building: Twelve, developed by Novare Group/Wood Partners
- September 2004—Atlantic Station redevelopment receives national Phoenix Award for "Best Brownfield Redevelopment"
- October 2004—Atlantic Station community announces Millennium Gate Project
- March 2005—Atlantic Station announces the grand opening date of The District: October 2005
- June 2005—IKEA, the Swedish home furnishing retailer, opens its first store in the Southeast in Atlantic Station
- July 2005—171 17th Street Building receives Silver certification in the Leadership in Energy and Environmental Design (LEED®) Core and Shell Development program. The office tower is the first LEED Silver-Core and Shell certified high-rise office building in the world and the first high-rise office building in Georgia to receive any LEED certification
- October 2005—The Atlantic Station redevelopment officially opens

Some aspects of the "greenness" of the development project include:

- Energy: An environmentally-friendly central cooling system will save building owners more than $35 million in construction costs, while operating more than 25 percent more efficiently than traditional building HVAC systems resulting in lower energy bills for tenants. A two-mile-long network of 36-inch pipes will deliver chilled water from a 50,000-square-foot central cooling plant to office, residential, and retail buildings.

- Recycled materials: During the property's reclamation, concrete building foundations were uncovered which were broken into smaller pieces and reused as backfill. This recycled concrete accounted for 132,000 cubic yards of material. Additionally, the 164,000 cubic yards of granite that was removed in order to create a level building site was crushed and reused as backfill. By using these large amounts of existing material, Atlantic Station, LLC reduced the amount of material that had to be taken to construction and debris landfills and lessened the material that had to be brought in from outside sources.

- LEED certification: The 171 17th Street building decided to pursue LEED certification. The 171 17th Street building is the first LEED Silver-Core and Shell certified high-rise office building in the world and the first high-rise office building in Georgia to receive any LEED certification.

- Transportation: The project includes pedestrian-friendly sidewalks on all surface streets in the development and priority parking and charging stations are available for electric vehicles.

Boisbriand, Quebec, Canada—Cherokee Investment Partners

Any discussion of brownfields redevelopment and greening these sites would be incomplete without a discussion of Cherokee Investment Partners (Cherokee) (www.cherokeefund.com) and one (or more) of their projects. First, an overview of Cherokee.

Cherokee is a private equity firm which has invested in the redevelopment of brownfield sites for over 20 years. Historically Cherokee focused only on the redevelopment of brownfield sites but over the past several years moved toward "building green."

The firm has always been identified as a leader in the remediation and redevelopment of brownfield sites and has now leveraged this leadership role into the sustainable redevelopment of properties after remediation.

An article on Cherokee and some comments from CEO Tom Darden provide insight on the history and strategy of Cherokee (http://money.cnn.com/2008/02/20/news/companies/gunther_cherokee.fortune/index.htm?postversion=2008022204).

According to the article, Cherokee launched its first private equity fund in 1996, raising money from institutional investors and high net worth individuals. Since 1996, Cherokee has raised more than $2.1 billion and remediated more than 500 sites in the United

States, Canada, and Europe. It has acquired properties such as Kmart stores in Canada, a General Mills flour mill in Vallejo, California, industrial properties formerly owned by United Technologies and Halliburton, and an abandoned Shell Oil refinery in Trieste, Italy. The view of Tom Darden is that "It's a contrarian strategy" and "We try to find value in something that's unpopular."

Darden's journey and that of Cherokee started when he returned to North Carolina to run his family's brick business. His leadership moved the business into sustainable business practices such as using waste sawdust instead of fossil fuels to power the firm's brick plants. This is where he developed the know how to remediate contaminated properties and ultimately moved beyond just remediation. In his words, "It's density versus sprawl" and "One of our goals is always to fill in the holes, rather than build out at the edges."

Figure 9.3 An aerial view of the Bosibriand site (from the Cherokee Project Description http://www.cherokeefund.com/trans/Boisbriand.pdf).

Figure 9.4 An illustration of a portion of the Cherokee Boisbriand development (from the Cherokee Project Description at http://www.cherokeefund.com/trans/Boisbriand.pdf).

Again, from the interview with Marc Gunther, "It's a fabulous time to be a buyer, and not a good time to have assets on the market," Darden says. "I'm like the guy with his head in the oven and his feet in the freezer. I feel about average."

Let's take a look at the Faubourg Boisbriand site north of Montreal as an example of the Cherokee approach (www.faubourg-boisbriand.com). The site is a former 232-acre General Motors manufacturing plant which will be redeveloped as a mixed-use site. The site required remediation of soil resulting from hydrocarbons from underground storage tanks and spills. The soil was remediated to residential cleanup standards.

Cherokee acquired the property in 2004 and brought the site through the entitlement process thereby moving the site closer to development as a mixed-use "lifestyle community." Significantly, the development plan was certified LEED Gold in the pilot project for the USGBC LEED Neighborhood Development (LEED-ND) program which was developed to integrate principles of smart growth, neighborhood design, and green building.

The development plan for the project includes the Boisbriand Town Center, 1400 residential units, a new community center, recreational facilities, and an urban square. A key aspect of the development is that the site is within walking distance of a proposed regional commuter rail line that will connect the site to downtown Montreal.

The development includes approximately 15 percent of the site dedicated to open space with walking and biking trails. Moreover, the plan calls for the planting of 2400 trees on the site and the reuse of 140,000 tons of concrete and 2.5 million square feet of asphalt.

As a testimony to the ambitious and thoughtful proposed plan the project was recognized by the Canadian Urban Institute's (CUI) 2007 "Brownie" Award for the best large-scale redevelopment project.

Alameda, California—Catellus

Catellus (www.catellus.com), similar to Cherokee, has built a business model focused on the redevelopment of brownfield sites and the greening of these sites.

The company was founded in 1984 as the Santa Fe Realty Corporation and conducted non-railroad real estate activities of the Santa Fe Southern Pacific Corporation (the merger of Santa Fe and Southern Pacific was denied but the realty corporation continued). The firm focused on industrial, mixed-use, and retail development projects. The company was renamed Catellus in 1990 and merged with ProLogis (www.prologis.com). Sustainability is a core business focus for ProLogis and they are members of the USGBC (www.usgbc.org) and the Chicago Climate Exchange (www.chicagoclimateexchange.com).

As part of their experience, Catellus recognizes the following key factors for project success:

- State (U.S.) support is essential
- Stakeholder input produces a high quality plan
- Need to embrace social and political objectives
- Public financing is a critical tool
- Incorporate sustainability principles.

One of Catellus's high visibility projects is the Alameda Landing site in the San Francisco Bay area, a former U.S. Navy industrial supply center. The 97-acre site is planned for mixed-use redevelopment, including residential and commercial elements. The waterfront development will retain some of the former warehouses as part of connecting the development to its history.

Alameda Landing will include 300 single family homes (including 25 percent as affordable housing), approximately 300,000 square feet of retail space, and up to 400,000 square feet of office space. The overall goal of the project is to create an interconnected community of residential, offices, retail stores, and open space adjacent to the waterfront within close proximity to San Francisco and Oakland. The development will build upon the history of the site and will include transforming the wharf into a community waterfront promenade. Building upon the interconnected aspects, the site will provide pedestrian and bicycle-friendly streets.

Alameda Landing will be readily accessible by car, BART, and water taxi. As planned, bus shuttles will also be built into the development plan to ensure easy access to and from the island.

The success of Catellus projects hinges on their robust community outreach and engagement. Elements of the community outreach by Catellus for this project included multiple open houses for the community, community workshops, and planning board workshops.

Treasure Island, California

Treasure Island (including the adjacent Yerba Buena Island) is a 450-acre brownfield site with a small existing community which is planned to be transformed into one of the most sustainable communities in San Francisco. While in the initial stages, the Treasure Island project in California holds great promise in being a unique development.

What the future may hold for the site was recognized by the awarding of the "Governor's Award for Environmental and Economic Leadership in Sustainable Communities" in November 2008. This recognition was awarded by the governor's office, The California EPA, and other state agencies to the City of San Francisco, Treasure Island Community Development (TICD), and community stakeholders. The award focused on the proposed green building and energy conservation features.

The review panel for the award referenced the project design as "exceptional" in its forward-thinking vision and commitment to environmental sustainability. According to Kofi Bonner, president of Lennar Urban's Bay Area Division and member of Treasure Island Community Development, "We thank the governor and those state officials who selected our development plan for this honor. The proposed redevelopment of Treasure Island truly reflects the environmental leadership displayed by San Franciscans and offers a model to other cities about how to build for the 21st century and beyond. We have been proud to work with the City of San Francisco, our partners, and the community to shape the future of a new Treasure Island."

In addition, Michael Cohen, City of San Francisco's Director of Economic and Workforce Development, has remarked that "Treasure Island will be one of the most environmentally sustainable large development projects in U.S. history."

The project is a public-private partnership between the Treasure Island Development Authority, a nonprofit public benefit corporation, and Treasure Island Community Development, which includes Lennar, Wilson Meany Sullivan, and Kenwood Investments.

The key elements of the project would include:

- Pedestrian friendly neighborhoods built around an intermodal transit hub will leave approximately two-thirds of the land available for a 300-acre park and open space.

- The green infrastructure proposed for the project will enable the development plan to strive for Gold certification under the LEED-ND program.

- A constructed wetland will provide wildlife habitat and serve a stormwater treatment function. A new wastewater treatment facility will recycle water for irrigation and other commercial uses.

- Approximately 6000 new homes are planned, with 30 percent considered as "affordable" (below market rates).

- New development will strive for the LEED NC environmental building standards. Buildings will be energy efficient and uses in the urban core will be served by an energy efficient central heating and cooling plant.

- Alternative energy sources such as solar and wind power will allow Treasure Island to create more energy than it uses during certain times of the day. Renewable grid-source power will be used for 100 percent of the development's power supply.

- A comprehensive transportation plan will prioritize walking and bicycling and maximizes use of public transit through congestion management, pricing, household transit passes, parking policies, and freeway ramp metering.

One hopes that this project is realized as it is a unique location, with a rich historical context and truly sustainable in its design.

South Waterfront Area—Portland, Oregon

The South Waterfront area of Portland was the former location of a power station, lumber mills, and scrap yards. The Portland Lumber Company was the first resident of the site, built in the 1880s, and it was used to generate electricity and steam as a source of heat for downtown Portland buildings. The plant closed in 1985 due to technological obsolescence and in the 1930s an electrical substation was built in close proximity to the lumber mill location, remaining in operation until 1989 when it was relocated to accommodate site redevelopment.

The site was successfully redeveloped into 480 residential units, 40,000 square feet of commercial space, a 74-room hotel, an athletic club, 26,500 square feet of retail and restaurant, an 83-slip marina, public breakwater, and 34 acres of public parks, streets, and open space. "Before and after" photos provide a sense of the scale of redevelopment (see Figure 9.5).

This project was considered progressive (by the USEPA) as it was a brownfields project before the term brownfields went mainstream. The Portland Development Commission (PDC) leveraged over 20 years of urban land development experience to the redevelopment of this brownfield site. The PDC and the Oregon Department of Environmental Quality (DEQ) collaborated to develop this mixed-use development project in downtown Portland.

As with many recent downtown redevelopment projects, the trigger for redevelopment was the demolition of a highway blocking public access from downtown to an adjacent river or bay (the San Francisco Embarcadero is a great example). The PDC committed to removal of the Harbor Highway in 1976 and in 1978 acquired the 73-acre site. The Portland City Council created the South Waterfront Redevelopment Program starting with 16 acres slated for redevelopment.

Figure 9.5 Aerial photo of Portland's South Waterfront site before 1979 (left) *(photograph by Portland Development Commission)*; aerial photo of the site after 2002 (right) *(photograph by Bergman Photographic Services, Inc.). Both photos courtesy of the PDC.*

The site required an evaluation of the nature and extent of soil and groundwater contamination. The remediation included "dig and haul" to address contaminated soil, the capping of areas to preclude exposure to contaminated soils, and innovative Willamette River bank stabilization procedures to eliminate the potential for sediment erosion. Remediation also included the five year operation of a groundwater monitoring program.

The project was also the first to enter the Oregon Voluntary Cleanup Program (VCP). Significant remediation challenges were the result of subsurface hydrogeology and the proximity to the Willamette River. The site contained high concentrations of lead and polynuclear aromatic hydrocarbons (PAHs) in subsurface soils and PAH contamination in the groundwater. Remediation planning required numerical groundwater modeling (due to proximity to the Willamette River) followed by a five year groundwater monitoring program.

The Oregon DEQ approved remediation plan consisted of the surface capping of discrete site areas, abandonment of a water intake structure, and installation of a riverbank stabilization system designed to provide access to the river and an enhanced habitat for migratory fish. Institutional controls, as one would expect, included prohibition of groundwater use, a surface cap maintenance program, and "specialized piling methods" and post-piling groundwater monitoring.

This transformation of the site was so successful it was the recipient of the prestigious Phoenix Award (www.phoenixawards. org), recognizing it as one of the best brownfield development projects in the U.S. at the time. The Phoenix Awards Institute, Inc., "an environmental and service nonprofit organization, is dedicated to honoring the groups that remediate and redevelop brownfield sites" and recognizes innovative solutions and revitalization projects that create "community assets." The award was established in 1997 and in 2000 became part of the U.S. Brownfields Conference sponsored by the USEPA, ICMA, and others.

The bulk of the project costs were incurred through private sector funding. However, the PDC contributed approximately $23 million to support the aggregation of properties and infrastructure development through Tax Increment Financing (TIF). In addition, a $350,000 grant was provided by National Resources Conservation Services (NRCS) and combined with TIF funding for waterfront bank stabilization and repair.

The site includes a four-acre South Waterfront Park and the Tom McCall Garden, which serves as both a neighborhood park and a landmark on the 60-mile urban Greenway Trail System that extends to both sides of the Willamette River. There is also an esplanade with stores and restaurants with an 83-slip marina that connects the existing waterfront park trail to the central city.

The site incorporates innovative energy-saving and water conservation techniques to minimize the neighborhood's impact on the environment, including:

- Bioswale filtration;
- LEED certified buildings; and
- An energy-efficient Portland Aerial Tram.

The South Waterfront area is also now considered safe for salmon (Salmon Safe) as a result of reduced stormwater runoff and reduced non-point source pollution (http://www.southwaterfront.com).

The economic benefits for the project include increased tax revenues (in 2002 tax revenues were over $1.5 million). And, as job creation is a critical issue these days, the project created over 440 jobs in professional, property and business management, maintenance, and the service industry.

One of the key success stories of the project was the demonstrated flexibility in regulatory engagement and "redevelopment-oriented solutions." The project is an excellent example of brownfields "done well."

Fulton Fish Market, Hunts Point—New York

The redevelopment of the Fulton Fish Market in New York received the Phoenix Award for excellence in brownfields development in 2006. The brownfields redevelopment project is highlighted because of how it fits into an overall community redevelopment project with a focus on incorporating sustainable development practices.

The original Fulton Fish Market was located in lower Manhattan and operated as a fish market for over 180 years. Ultimately it became the largest fish market in the U.S. and outgrew its Manhattan location.

The market is being relocated to the Hunts Point site in the South Bronx and will be part of an overall redevelopment plan—the "Hunts Point Vision Plan," which is a comprehensive framework for promoting a competitive business environment and a sustainable community in this area of the Bronx.

The "vision plan" incorporates elements of sustainability and the creation of new jobs with the goal of establishing a viable economic base through industrial development and a supporting residential community. The City of New York invested approximately $27 million to get the project moving forward.

The relocation of the fish market took almost 40 years by the city and State of New York to consolidate the city's food distribution enterprises in Hunts Point. The Fish Market's move further establishes Hunts Point as the leading food distribution center in the country.

Redevelopment of the site included the remediation of a coal gas production site with waste and contaminants such as: waste wood chips, petroleum-based contaminants, purifier waste, and coal tar. Over 28,000 tons of coal tar and 8000 tons of purifier waste were removed from the site. The waste material was trucked to American Refuel in Buffalo, NY where it generated energy for steam and electricity producing over 7.6 megawatt hours of electricity for the Buffalo grid—enough to heat 10,000 homes for three months.

The total cost of the project was estimated at approximately $85 million with remediation costs at $12 million (http://www.nyc.gov/html/oec/html/brown/brownnews.shtml).

Homebush Bay—Sydney, Australia

One of the most ambitious brownfield projects built green was the Homebush Bay Olympic site in Sydney. The project was (and remains) a breakthrough due to the extent of contamination and the degree of green practices embedded into the project.

The site ultimately became a showcase for incorporating ecologically sustainable development (ESD). ESD was defined in Australia in 1992 as a National Strategy for Ecologically Sustainable Development providing a framework for sustainable principles for redevelopment in the bid for the 2000 Olympic Games.

According to the Olympic Coordination Authority (OCA) the Homebush site was a former brick works, a naval armaments depot, and a general dumping ground for waste which required significant remediation (http://www.sydneyolympicpark.com.au/__data/assets/pdf_file/0003/849/towards_sustainable_urban_ecology_olympic_site_message.pdf).

The City of Sydney's environmental commitments and recommendations for future Olympic host cities were outlined in the "Environmental Guidelines for the Summer Olympic Games, September 1993."

The key principles outlined in this document include:

- Energy conservation
- Water conservation
- Waste avoidance and minimization
- Protection of human health
- Protection of natural and cultural heritage.

The process of translating these key principles fell to the Australian OCA that was ultimately responsible for the construction activities and the sustainable development practices. Key steps in-

cluded (http://www.sydneyolympicpark.com.au/__data/assets/
pdf_file/0003/849/towards_sustainable_urban_ecology_olympic_
site_message.pdf):

1. Determine what was at the site. The primary step was to determine the nature and extent of contamination and to the extent possible preserve the current ecology of the site.

2. Provide strategies to utilize the site attributes. Preliminary concepts for site development were outlined within the context of preserving a sensitive environmental setting.

3. Provide guidance to site managers and workers. This was where the "rubber met the road" and required providing actual guidance of the developers, designers, planners, and managers in the implementation of ESD principles.

   The key ESD performance areas were:

   • Conservation of species—flora and fauna, people, and their environment.
   • Conservation of resources—water, energy, construction materials, open space, and topsoil.
   • Pollution control—air, noise, water, light, soil, and waste management.

4. Prepare the tools for ESD implementation. The guidance was then translated into tools that could be used by a variety of project managers. The "ESD tools" included:

   • "Environmental tender specifications (ETS) for inclusion in all OCA contracts to allow companies applying for the work to demonstrate how they will comply with OCA's environmental commitments.
   • An environmental management system (EMS) to outline OCA's environmental commitments and environmental legislative requirements at every stage in all projects (design, construction, and operations).
   • Environmental management plans (EMPs) for design, construction, project management, and operation of venues. These specified the controls in place to manage environmental risks and impacts arising from tasks and identify how ESD was being specifically implemented into each project.

- Consultation—Expert technical panels were established to advise on specific research findings and their potential application by OCA on construction materials, ecology, energy, landscape and open space, waste management, and water/wastewater.

- The Olympic Environment Forum was established that included representatives from the New South Wales (NSW) Environment Protection Authority and green groups to discuss environmental issues associated with the Olympic Games.

- A community stakeholder working group, Homebush Bay Environment Reference Group, was established to discuss contamination and remediation issues and develop community and education projects.

- Environmental training was conducted to ensure all staff and contractors received appropriate environmental training that increased environmental awareness and ensured participants were aware of OCA's environmental commitments, NSW environmental legislation requirements, and site-specific issues.

- Auditing and monitoring programs were set up to ensure the tasks put in to achieve ESD were being undertaken and achieving what they had set out to do."

The former brownfield site is now an international showcase for how sustainability practices are embedded into a mixed-use development project. The site has outstanding sporting venues, extensive open space, and residential developments.

Key sustainability practices incorporated into the Homebush site include the following five elements (http://www.sydneyolympic-park.com.au/__data/assets/pdf_file/0003/849/towards_sustainable_urban_ecology_olympic_site_message.pdf).

Energy Conservation

Energy conservation was achieved through a diverse range of energy efficient-building designs and technologies. These include the use of energy from renewable sources ("green power"), solar energy, and designs that conserve energy, including skylights and the orientation of buildings to maximize natural heating, lighting, and ventilation.

- Nineteen towers (the Towers of Power) housing solar PV panels and providing shade, seating, and lighting line Olympic Boulevard. The energy collected in the solar cells is supplied to the national grid and generates approximately 160,000 kilowatt hours of energy a year, or enough power for 38 Australian households per year.
- The Athletes Village is one of the world's largest solar-powered suburbs, with photovoltaic solar energy panels fitted to 665 homes.
- The Novotel Hotel at Sydney Olympic Park has 400 square metres of solar hot water collectors, which was the largest solar hot water system in the southern hemisphere at the time. The hotel also achieves approximately 40 percent energy efficiency compared to comparable hotel designs.
- At Stadium Australia, two 500 kilowatt gas co-generation engines supply electricity and heat to meet a large share of the venue's energy requirements. These produce about 40 percent less greenhouse gases and solar PV installed on the roof of the Superdome provide energy back to the grid.
- At the Sydney Aquatic Centre air conditioning was designed to cool only the air immediately surrounding the spectators which translates into less energy required for both cooling the venue and heating the pool.

Water Conservation

Water is conserved throughout the development through water efficiency and recycling practices. Water efficient appliances coupled with the use of native trees and shrubs reduce water needs and stormwater is collected from the roofs of Stadium Australia and the Sydney Showground to provide water for irrigation.

In addition, the most significant water conservation initiative was the Water Reclamation and Management Scheme (WRAMS). Sewage generated from Olympic venues and Newington Village plus stormwater from the site is collected, treated, and cleansed for non-drinking water uses such as watering plants and toilet flushing. Stormwater is collected and stored in water quality control ponds and tested before use.

Waste Reduction and Pollution Control

During the building of the development construction materials were reused and recycled. This resulted in the recycling of 40,000 cubic metres of earth excavated during construction that was used to create embankments at the Athletic Center and 94.7 percent of construction waste at the Sydney Showground was recycled.

Protection of Human Health
Based upon stakeholder engagement there was a preference for the treatment and containment of waste onsite rather than removal to other areas. The strategy chosen to prevent potential health impacts was "consolidate, contain, and cap" plus further treatment of some waste. Where possible, additional treatment included biodegradation of petroleum of contaminants and thermal treatment for other contaminants. The waste "consolidation mounds" are capped with compacted clay and topsoil and a contaminated groundwater collection system is installed. Over $137 million was spent on remediating the site.

Conservation of Species
The site is home to a number of threatened species and includes a significant diversity of species for an urban landscape. The Green and Golden Bell Frog (an endangered species in NSW) was discovered at the site and a management plan to ensure its survival was developed. The frog is now flourishing in a range of areas. Birds protected under international treaties visit the site, and studies were implemented to monitor the type of species present and protect their habitat. Important saltmarsh areas were also preserved on the site as part of the conservation focus during development.

Overall Environmental Accomplishments
The achievements of the Homebush Bay site were independently evaluated by the Earth Council. The OCD learned valuable lessons for potential application at other sites, including:

- Planning and construction,
- Energy and water conservation,
- Waste avoidance and resource conservation,
- Protection of air, soil, and water quality through remediation, and protection of natural and cultural heritage.

In many ways the Homebush Bay Olympic site set the stage for future Olympic venues and raised the bar on "green expectations."

The Olympic project now to watch is Vancouver. They won a CERES award for Commendation for Innovative Reporting (Vancouver Organizing Committee of the 2010 Olympic and Paralympic Winter Games). The Vancouver Organizing Committee's Sustainability Report is serving as a "planning tool for making the 2010 Olympic Games more sustainable and is a clear example for how other project-based entities can and should address sustainability."

China

The climate for sustainable development projects in China is changing. There are two noteworthy projects worth mentioning to illustrate approaches currently used and where projects may be headed.

In 2007 two projects in China were awarded "top honors" by the Waterfront Center. The Waterfront Center is a nonprofit organization which has promoted waterfront design and planning for 25 years.

The two projects reflect the balance of development with environmental protection and in some cases "enhancement." The two projects were highlighted in the paper, "Brownfields and Redevelopment in China: Learning from the American Experience and Two Award Winning Redevelopment Projects in China" by Barry Hersh of the Real Estate Institute.

The Floating Gardens, Yonging River Park, Taizhou City

The project was the redevelopment of a former channelized concrete riverbank. The redevelopment included a wetland and park which meet flood control requirements and also provide open space and cultural amenities in an urban setting. The park consists of essentially two layers: a natural layer uses the flood pattern to reconstruct a wetland system with an outer system of ponds within the body of the park. Both systems flood with the outer wetland that remain submerged during the dry season. The presence of dense wetland plants, "floating" trees, and planted bamboos combine well with the river and the urban setting.

Hong Kong Wetland Park

The 61-hectare freshwater wetlands and tidal wetland is located in a densely populated suburb of Hong Kong. The Wetland Park is designed to clean stormwater and also provides an "outdoor classroom" to promote wetland plant ecology. The park combines educational exhibits and laboratory spaces. The park is now viewed as an example of China's capacity to incorporate sustainable design elements and build upon well established practices.

REVOLUTIONARY GREEN

When examining truly revolutionary projects I believe we need to start with the BedZED project in England and One Planet Living Principles® (www.oneplanetliving.org). My rationale for this is that the BedZED project was (and in some ways still is) groundbreaking.

Also, as part of a revolutionary green portfolio of projects, we will look at the partnerships that were crafted to move these unique projects forward. Increasingly, stakeholder partnerships (with stakeholders who are not what you would expect) are an integral part of a successful project.

First, let's provide an overview of One Planet Living® (OPL) and its key principles. Essentially, the overall goal of OPL is to promote principles of sustainable development by establishing sustainable communities globally. The belief is that by establishing these global communities the program will have to transform the surrounding regions and influence public policy.

The 10 guiding principles of OPL are outlined in Table 9.1.

Now, let's examine how these principles were applied at the BedZED and other OPL projects.

BedZED

BedZED stands for the Beddington Zero Energy Development located near Wallington, England in the London Borough of Sutton. The project was a partnership between:

- BioRegional (www.bioregional.com);
- Bill Dunster Architects;
- Peabody Trust;
- Arup; and
- Gardiner and Theobald.

BioRegional is a not-for-profit, founded in 1992, focused on the goal of "inventing and delivering practical sustainability solutions." The core beliefs of the organization are to create "a future in which

Table 9.1 The 10 guiding principles of OPL (from the OPL Web page, www.oneplanetliving.com). *continued on next page*

GLOBAL CHALLENGE	OPL PRINCIPLE	OPL GOAL AND STRATEGY
Climate change due to human-induced buildup of carbon dioxide (CO_2) in the atmosphere	Zero Carbon	*Achieve net CO_2 emissions of zero from OPL projects* Implement energy efficiency in buildings and infrastructure; supply energy from onsite renewable sources, topped up by new off site renewable supply where necessary.
Waste from discarded products and packaging creates a huge disposal challenge while squandering valuable resources	Zero Waste	*Eliminate waste flows to landfill and for incineration* Reduce waste generation through improved design; encourage reuse, recycling, and composting; generate energy from waste cleanly; eliminate the concept of waste as a part of a resource efficient society.
Travel by car and airplane can cause climate change, air and noise pollution, and congestion	Sustainable Transport	*Reduce reliance on private vehicles and achieve major reductions of CO_2 emissions from transport* Provide transport systems and infrastructure that reduce dependence on fossil fuel use, e.g., by cars and airplanes. Offset carbon emissions from air travel and perhaps car travel.
Destructive patterns of resource exploitation and use of non-local materials in construction and manufacture increase environmental harm and reduce gains to the local economy	Local and Sustainable Materials	*Transform materials supply to the point where it has a net positive impact on the environment and local economy* Where possible, use local, reclaimed, renewable, and recycled materials in construction and products, which minimizes transport emissions, spurs investment in local natural resource stocks, and boosts the local economy.
Industrial agriculture produces food of uncertain quality and harms local ecosystems, while consumption of non-local food imposes high transport impacts	Local and Sustainable Food	*Transform food supply to the point where it has a net positive impact on the environment, local economy, and people's well-being* Support local and low-impact food production that provides healthy, quality food while boosting the local economy in an environmentally beneficial matter; showcase examples of low-impact packaging, processing, and disposal; highlight benefits of a low-impact diet.
Local supplies of fresh water are often insufficient to meet human needs due to pollution, disruption of hydrological cycles, and depletion of existing stocks	Sustainable Water	*Achieve a positive impact on local water resources and supply* Implement water use efficiency measures, reuse and recycling; minimize water extraction and pollution; foster sustainable water and sewage management in the landscape; restore natural water cycles.
Loss of biodiversity and habitats due to development in natural areas and over exploitation of natural resources	Natural Habitats and Wildlife	*Regenerate degraded environments and halt biodiversity loss* Protect or regenerate existing natural environments and the habitats they provide to fauna and flora; create new habitats.

Table 9.1   The 10 guiding principles of OPL (from the OPL Web page, www.oneplanetliving.com). *continued from previous page*

GLOBAL CHALLENGE	OPL PRINCIPLE	OPL GOAL AND STRATEGY
Local cultural heritage is being lost throughout the world due to globalization, resulting in a loss of local identity and wisdom	Culture and Heritage	*Protect and build on local cultural heritage and diversity* Celebrate and revive cultural heritage and the sense of local and regional identity; choose structures and systems that build on this heritage; foster a new culture of sustainability.
Some in the industrialized world live in relative poverty, while many in the developing world cannot meet their basic needs from what they produce or sell	Equity and Fair Trade	*Ensure that the OPL project's impact on surrounding communities is positive* Promote equity and fair trading relationships to ensure the OPL community has a beneficial impact on other communities both locally and globally, notably disadvantaged communities.
Rising wealth and greater health and happiness increasingly diverge, raising questions about the true basis of well-being and contentment	Health and Happiness	*Increase health and quality of life of OPL project members and others* Promote healthy lifestyles and physical, mental, and spiritual well-being through well-designed structures and community engagement measures, as well as by delivering on social and environmental targets.

everyone can enjoy a high quality of life, while living within their fair share of the Earth's resources and leaving space for wildlife and wilderness."

BioRegional was founded by Sue Riddlestone and Pooran Desai OBE, and began life in the Sutton Ecology Centre. Initial projects focused on environmental issues such as closed loop recycling and biomass energy projects.

These projects were the basis for expansion by the organization in the late 1990s resulting in the successful BedZED project through the partnership with OPL (along with the World Wildlife Fund) and establishing offices in China, South Africa, and Canada.

Bioregional is also engaged in other "environmental projects" such as (www.bioregional.com):

- "OPL communities, planned for every continent by 2012;

- B&Q One Planet Home, developing a sustainability action plan for the retailer and a customer-focused campaign and product range;

- The Laundry, a closed loop recycling social enterprise for small businesses in central London;

- One Planet Products, a sustainable construction materials buying club;

- HomeGrown Charcoal, produced through a decentralized network of charcoal burners to reduce transport CO_2 by 90 percent; and

- BioRegional MiniMills, developing technology to make paper pulp from straw and recover energy from the effluents."

The other not-for-profit partner in the BedZED project was the Peabody Trust (www.peabody.org). The Peabody Trust is one of London's oldest housing associations which (according to its Web page) "exists to tackle poverty, provide good, affordable housing and to make a difference through every project or initiative it undertakes."

The Peabody Donation Fund was established in 1862 by George Peabody which was incorporated by the English Parliament in 1948. The Trust (as of 2005) owns over 19,000 properties spread across London which house almost 50,000 people. The Trust now focuses on sustainable housing as part of its overall strategy.

The mixed use BedZED project was built between 2000 and 2002 and has gained worldwide recognition and was shortlisted for the Stirling Prize awarded annually by the Royal Institute of British Architects. The project is a development of "flats, maisonettes, and town houses" and approximately 2500 m² of workspace/office and community accommodation including a health centre, nursery, organic café/shop, and sports club house.

The BedZED project could have been built in any urban setting and conceivably on a greenfield site (however, it would have contradicted the vision of the project partners). Instead it was built on a brownfield site in the London borough of Sutton (see Figure 9.6).

The project is notable for several reasons; it is the UK's largest "eco-village" and it is arguably the best example of not only sustainable development but of "sustainable living" in the UK.

A few statistics and attributes of the project (www.bio regional.com):

- Zero-energy—The project only uses energy from renewable sources generated onsite. In addition to solar panels, tree waste

Figure 9.6 Photo of the BedZED mixed-use development in Wallington, South London, UK. (http://www.paulmiller.org/bedzed.jpg)

fuels the development's heat and electricity. The heating requirements of BedZED homes are around 10 percent of a typical home. The super insulated and wind driven ventilation system incorporates heat recovery, and passive solar gain is stored in thermally massive floors and walls (which reduces the need for both electricity and heat).

- High quality—The apartments are finished to a high standard to attract the urban professional.

- Energy efficient—The houses face south to take advantage of solar gain and windows are triple glazed and have high thermal insulation.

- Water efficient—Most rainwater falling on the site is collected and reused. Appliances are chosen to be water efficient and use recycled water where possible.

- Low impact materials—Building materials are selected from renewable or recycled sources located within a 35 mile radius of the site to minimize the energy required for transportation.
- Waste recycling—Refuse collection facilities are designed to support recycling.
- Transport—Car parking spaces are limited, but residents share in a car pooling program.

The buildings have brightly colored ventilation cowls that move with the breeze and vastly improve the movement of air through the units. In addition, photo voltaic panels adorn the roof tops providing energy to the units.

All of the BedZED's units face south, maximizing the sun's heat and light and, as previously discussed, are highly insulated to retain heat in the winter. Each resident has a small private garden and interior fixtures are efficient with low energy lighting and water saving appliances. Estimates by BedZED indicate that water-efficient appliances have cut the development's main water use (the public supply) to 91 liters per person per day, compared with a UK average of 150 liters.

The BedZED project is a "third generation design" developed over a five year period by Bill Dunster architects and their ZEDfactory, Ltd practice (www.zedfactory.com). According to ZEDfactory, BedZED was initially based upon the Hope House at Hampton Court. This next generation design provides a balanced approach to the needs of "residents, small and large businesses, a healthy living centre, a nursery, a café/telecommuting centre, the need for sunlight and daylight, an economic construction system and high levels of insulation without losing contact with the outside world" coupled with the objectives of the Peabody Trust to build a mixed-use project on a brownfield site. The BedZED project, while revolutionary, has not been without operating problems. This is not surprising in any ambitious venture such as this one.

Candidly, Bill Dunster has acknowledged that "We don't claim it's a perfect project" and "There's no doubt it can be done better." (http://www.chinadialogue.net/article/show/single/en/495-An-architect-s-sustainable-dreams.)

BedZED was designed to generate all of its energy from renewable sources such as solar PV, wind, and biomass. However, the development has experienced problems with its combined heat and power (CHP) plant resulting in a greater need to rely on domestic hot-water tanks, which double as radiators. "The combined heat and power plant has been very problematic," Dunster acknowledges. "And that's a big lesson learned. The higher the technology, the quicker it breaks and the quicker it becomes useless. The higher the maintenance, the more uneconomic it is to run. So, luckily—seven years later—there is a replacement coming for that, which will be reliable and need less maintenance. Technology does move on."

This has resulted in fluctuations in the renewable energy output between meeting 80 percent of the site's needs in 2003 and 11 percent more recently (2006). The project has also had to rely on public water supply when rainfall was less than estimated.

Moreover, the onsite wastewater treatment system, a "Living Machine" which naturally uses plants to treat wastewater, has been off line. "The 'Living Machine' has worked but has not been maintained," he adds, "and has been difficult to be adopted by the water authorities, who have been reluctant to engage in micro-treatment of water onsite—as the electric utilities have been reluctant to engage in electric and heat generation onsite."

Whatever the problems have been with some aspects of the sustainable design, the energy "load reduction" design elements have worked as one would expect. According to Dunster, "The (energy) load-reduction exercises have basically worked" and "the buildings don't need much heat."

Dunster's opinion of performance challenges is shared by others. Comments by Ms. Bennie of the Peabody Trust in 2005 indicated "fundamental" problems with "contaminated" run-off water from green roofs, underperforming photovoltaic cells, and a lack of privacy for residents. Some of the criticisms include that design was considered second behind environmental concerns such as daylighting, and design problems such as an apparent loss of privacy and a desire by some residents for increased parking space. Other concerns raised by Bennie included green roof performance (too much runoff) and PV performance.

It is worth noting that while the overall design works well, attempts to change some aspects of the residents' behavior have been more challenging. The project encourages telecommuting, recycling, car pooling, and the use of a local organic food delivery service. These services and practices have been slow to take. As with any "green" practice, behavior is slow to change and BedZED is no exception.

Implications of BedZED and Scaling the Project

Based upon the performance of the BedZED project it would potentially be possible to reduce the urban sprawl in the UK by approximately 25 percent. This is an impressive statistic with significant implications, not just for the UK, but globally.

According to a 2006 interview by Maryann Bird with Bill Dunster titled "An architect's sustainable dreams" (http://www.chinadialogue.net/article/show/single/en/495-An-architect-s-sustainable-dreams) the BedZED lessons are being rolled out on a global basis.

One of Dunster's key areas of interest is the "microgeneration" of energy pioneered at BedZED. Dunster believes that microgeneration at the scale of 2000 homes or so would be revolutionary. According to Dunster, "The key is reducing the cost—and that can only be achieved by collaborating with people like the Chinese and helping them get the economies of scale in their industries so that the costs come down, so that we can adopt these concepts in Europe as well . . . So it's a double carbon-save, a double carbon-win."

"We're trying very hard to get our supply chain, all our components, made in China now," he says. "The Chinese can't afford to import any [energy microgeneration] products at all . . . The Chinese market is very cost-conscious. All these ideas rely on keeping things simple and reducing the costs. So we've got to be, all the time, designing with economies of manufacturing in mind."

"In our company now, we're no longer [only] architects," he adds. "We design our own low-energy components, and we try to help all the way through the process, rather than just saying, 'I've designed this—go and make it.'"

As one would expect, one of Dunster's offices is working on solar-powered air conditioning and dehumidification specifically for the Chinese market.

"But the barriers in both Europe and China are always the same," he finds. "You can't afford to do something different because nobody else is doing it. So all the time you're building prototypes which are too expensive. It's a vicious circle."

Dunster is also working on BedZED designs in China: a BedZED-sized low-rise development on the outskirts of Beijing and a high-density project in Changsha, in south-central China's subtropical Hunan province. Other projects consist of a building in the Changsha project, in the suburban district of Kaifu, consisting of flats, retail outlets, and a hotel.

The concept is that this is "a prototype for a new kind of urban block, replacing tower blocks and Wal-Marts, which is what they do now" in China. The site is "all mixed up—work, retail, and community spaces mixed with houses" and "what's important is that air blows through the east- and west-facing (rather than south-facing) homes. They have a cooling problem, not a heating one." Because the Changsha site calls for ventilation, fresh air and dehumidification, passive, horizontal wind cowls are incorporated in the design."

The other "Dunster project" is a zero (fossil) energy farm (ZEF), about 30 miles from Bejing, a new community beyond the Chinese capital's last ring road and surrounded by fields. In the early planning stages, it is designed for microgeneration, with solar electricity, a wind turbine, and a reed bed for water-filtering on top of a community building. The development is to include varying sizes of homes, a fishing lake, a farm shop supported by local agriculture, a café, work space, and underground parking in the high-density part of the site. A show house is nearly complete. "We think this is significantly more advanced than the stuff that's being built in the UK," he says.

"All of the microgeneration technology is allowing decentralized energy production and reimpowering ordinary people," Dunster points out. "We think this is the future, because there is a direct relationship between having your own set of solar panels, your own micro-wind turbine, and actually being aware of your carbon footprint and energy use. The feedback is very direct."

Dunster's view of the "smart grid" is very insightful. He believes that there is not enough renewable energy to meet current demand. He (correctly, in my opinion) favors energy efficiency and microgeneration onsite to reduce demand.

It is also interesting to note that when BedZED was built, climate change was not top of mind as it is now with the public and private industry. Dunster believes that, "When we built BedZED, we didn't see climate change coming so soon. If we had, we would have passive cooling, more shading, solar thermal collectors, micro-wind turbines. No CHP, but a wood-chip boiler instead. And we'd try and get people, the residents, to be fully engaged in the design process. That would sort out the hostility between the uber-greens and the people who just want to get on with it."

The OPL Mata de Sesimbra Project in Portugal

The Mata de Sesimbra in Portugal (see Figures 9.7 and 9.8) is another example of a revolutionary OPL supported project (Facts and

Figures about Mata de Sesimbra [doc, 222 KB], Deutsche Welle Radio reporting on the announcement of the Mata de Sesimbra project [mp3, 6.06 MB]).

The Portuguese developer Pelicano is developing a €1.1 billion project to build 5000 zero-carbon, zero-waste homes, hotels, and shops using OPL principles. The project will not include the restoration of an industrial brownfield site, but a "nature-restoration" site with approximately 5000 hectares of surrounding land being returned to native Mediterranean woodland after years of degradation from logging plantations.

The project is at the intersection of restoration, sustainable development, and job creation. The project is expected to create approximately 11,000 jobs and the sales of the properties will fund work to save endangered species throughout Portugal. The project will also include a "sustainable public transport" system estimated at €90 million, along with car pooling and hybrid shuttles.

The key initial design elements of the project are outlined below. Bear in mind that these are goals and the final performance of the project may be quite different. According to the project announcements and specifications (Facts and Figures about Mata de Sesimbra, Deutsche Welle Radio reporting on the announcement of the Mata de Sesimbra project):

SUSTAINABLE MATERIALS

- At least 50 percent use of recycled materials such as cement
- At least 90 percent elimination of toxic materials
- At least 50 percent of materials from a 50 km radius
- At least 30 percent reduction of embodied CO_2 in construction, transport, and use of materials

ZERO CARBON

- Reduce energy use of lights and appliances by 40 percent by installation of A-rated appliances and designing buildings to maximise daylight
- Reduce ventilation by 44 percent through the use of natural "wind-driven solutions"
- Reduce water heating energy use by 60 percent through the use of flow restrictors, spray taps, efficient shower fittings, and use of solar water heating

- Reduce space heat/cooling by 95 percent through the use of passive solar heat, glazing and insulation, and south-facing facades and shading solutions

ZERO WASTE

- Reduce waste production by 25 percent through the elimination and reuse of food packaging, promoting reusable packaging, and discriminating against disposable products
- Ensure at least 25 percent of waste is recycled, and 0 percent incinerated
- Compost over 90 percent of organic waste
- Reduce landfill to 5 percent of national average

RENEWABLE ENERGY AND ENERGY EFFICIENCY

- Achieve 100 percent renewable, non-fossil energy production by including photo-voltaics in architectural design, solar thermal design, small-scale biomass heating, and water ponds for space cool system

LOCAL RESOURCES

- At least 25 percent of food consumed within the development to come from 50 km radius, by organizing a network of local product and service suppliers
- At least 15 percent of key services provided locally, by offering preferential contracts for purchasing and hiring of local services, and supporting the development of those not currently available
- Employ 6000 people from local area in order to reduce unemployment, and develop skills enhancement programs

SUSTAINABLE TRANSPORT

- Reduce car use to 0 percent in urbanized areas through hard barriers and promoting alternatives such as walking and cycling
- Implement €100 million public transport and road network
- Increase average number of passengers per car to three per vehicle through car pooling programs

FAUNA & FLORA CONSERVATION

- Convert existing degraded monoculture forest, consisting of 90 percent exotic eucalyptus and pine, to native oak woodlands

Figure 9.7   Conceptual design for one of the Mata de Sesimbra housing units.
(http://www.bioregional.com/oneplanet/Sesimbra)

- Implement a €20 million conservation strategy
- Create local habitat corridors, and connect to regional ones; and conserve and increase sensitive wetlands and dunes
- Conserve and improve conservation status of target and vulnerable species, such as Bonelli's Eagle; and protect and regenerate pockets of biodiversity (such as Cork forest and Mediterranean maquis), and important habitats (such as bird of prey nesting sites) from disturbance

WATER CONSERVATION

- Reduce overall water consumption by 25 percent through the use of gray water recycling, low-flush systems, low-flow taps, and rainwater catchment

- Apply EGA Ecology Unit/Audubon Society rules to achieve major savings in sports and leisure-related water consumption

QUALITY OF LIFE PLAN

- Apply environmental quality indicators, such as air and noise pollution
- Build a €100 million program of sports, leisure, cultural, and educational facilities
- Provide minimum area of social facilities, open spaces, and green areas per capita
- Achieve positive values for attitudes to facilities and the community, and negative values for stress levels, through communication and participation programs

HERITAGE PLAN

- Preservation and educational programs, and supported programs to promote local heritage through strategy to promote natural, cultural, and historic values

Figure 9.8 Conceptual design for a portion of the Mata de Sesimbra development project. (http://www.bioregional.com/programme_projects/opl_prog/portugal/mds3_lrg_image.htm)

- Advertising of local heritage, facilities, and products such as crafts
- Promote OPL awareness through a "One Planet Living Centre"

As of November 2008, BioRegional and WWF officially presented their proposals to community stakeholders for input. The One Planet Living Information Center opened in March 2009 in Sesimbra and stakeholders, including local NGOs and businesses, are being invited to participate in a "Mata de Sesimbra Forum" to solicit input on the project (http://www.bioregional.com/one-planet/Sesimbra).

As one would expect from an innovative project of this size, stakeholder engagement will be critical in ensuring that the project is not only successfully designed and built but also supports the local communities and economy. A sustainable project needs to meet the "test of time" and engagement can contribute to any long term success.

Sonoma Mountain Village

I was fortunate to have visited the Sonoma Mountain Village site and the Sonoma Mountain Business Cluster (SMBC) in early 2009 to present on the topic, "Sustainability as a Driver for Innovation." This brief visit provided a wonderful glimpse into a planned OPL project (the first North American OPL site) and a co-located successful business cluster (see Figure 9.9).

Briefly, the SMBC is a nonprofit business incubator with a focus on developing start- up businesses engaged in sustainability (environmental and social performance) technologies. The incubator resides in a 29,000 square foot LEED Gold level standard facility. The SMBC provides support in the areas of (http://sonomamountainbusiness-cluster.com):

- "Mentoring and coaching;
- A service network;
- An intellectual and entrepreneurial environment;
- Educational and networking events;
- Contact to investment capital; and
- An affordable full-featured physical infrastructure"

The mission of the SMBC is to:

- "Stimulate regional economic development through entrepreneurship and enterprise development;

- Reinvigorate Sonoma County's technology industry making it a recognized center for sustainable resources and socially-relevant technologies;

- Provide infrastructure, training, management assistance, financial resources, and support networks that enable new and emerging companies to succeed and prosper, creating value-added jobs for the local region through the commercialization of technology; and

- Create a hub activity contributing to the development of a thriving live/work community in the eco-conscious Sonoma Mountain Village."

I bring up the business cluster because I view this as such an important part of what I believe will contribute to the success of the Sonoma Mountain Village. Developing viable (read: profitable) new businesses with a sustainability mission that will provide the economic platform to sustain a community is critical. The SMBC is off to a great start.

So let's take a look at the only North American OPL project, Sonoma Mountain Village.

While this site is not a brownfield, the site itself is a former Agilent corporate high-tech campus. The Agilent facility at one time provided approximately 3000 jobs which will be more than replaced by redevelopment.

Part of the vision for the development (and the SMBC) is to create new jobs and to physically recycle much of the former Agilent facility. The site is located in a physically beautiful area with numerous open spaces and parks. The Village will enhance this setting through building in the new urbanism style (also known as traditional neighborhood design, "neotraditional" neighborhood design, and transit-oriented development).

New urbanism embraces a wide range of housing types and is designed to be "walkable" or in my mind "human scale." The essential aspects of new urbanism are:

- Neighborhoods should be diverse in use and population;

Figure 9.9 Current design illustration for Sonoma Mountain Village.
(http://www.sonomamountainvillage.com/illustrations.htm)

- Communities should be designed for the pedestrian and transit as well as the car;
- Communities should be shaped by physically defined and universally accessible public spaces and community institutions; and
- Urban spaces should be framed by architecture and landscape design that celebrate local history, climate, ecology, and building practice.

The Village also embraces design principles such as smart growth and smart code. Smart growth principles state that planning and transportation should be concentrated at the "city center" to avoid development sprawl. The principles are designed to promote transit oriented and walkable development.

Smart code, as the name implies, is a holistic approach to planning and zoning. Basically it is designed to translate smart growth principles into the creation of walkable neighborhoods and mixed-use development (http://www.sonomamountainvillage.com/overview.htm)

In addition to making OPL come to life, the Village is a member of the USEPA Climate Leaders program and the USGBC. The Village is striving for LEED-ND Platinum certification.

The key specifications for the Village are:

- Starting in late 2009 the build out will span 12 years;
- 200 acres in area;
- $1 billion investment;
- 1900 new homes;
- Creation of a projected 3000 new jobs; and
- 500,000 square feet of commercial, retail, and office space.

There are currently 21 businesses within the Village, including AT&T, Comcast, Codding Construction, and Codding Steel Frame Solutions.

The Village design calls for abundant open space along with over 25 acres of parks, a community center, walking trails, an all weather soccer field, fitness center, and a "Lifelong Learning Institute."

The project recently won the (California) 2008 Governor's Environmental and Economic Leadership Award. (http://www.calepa.ca.gov/awards/geela/2008/WinnerSummary.pdf)

As outlined in the award document, Jake Mackenzie, the mayor of Rohnert Park commented "Codding Enterprises' transformation from a mainstream developer to a deeply sustainable mixed-use developer focused on a restorative impact deserves encouragement from all responsible organizations."

Via Verde Bronx Development

I am keenly interested in brownfield sites that are not only built green but create "green affordable housing." The two leaders in this area within the U.S. are Jonathan Rose and Enterprise Community Partners (Enterprise). A brief look at two projects from Jonathan Rose and Enterprise will provide some insight as to the transformative impact of this strategy.

The first site in the South Bronx, New York is a vacant 60,000 square foot lot at Brook Avenue and East 156th Street which was condemned by the City of New York in 1972. The site is a "classic" New York City brownfield site with abandoned railroad tracks and "likely" contaminated soil and groundwater.

The proposed project is a 202-unit apartment complex, including an 18-story tower, mid-rise building with duplex apartments and townhouses. Sixty-three units would be co-op apartments for sale, with the remaining units as rentals. The Via Verde project is planned to meet at least LEED Gold level and the Enterprise Green Community standards (www.greencommunitiesonline.org).

According to Shaun Donovan, the New York commissioner of the Department of Housing Preservation and Development, "We started out on this process to try to raise the level of design and the level of sustainability for housing not just on this site but with the hope that this could be really a model . . . Given the responses we got, I think there's a real opportunity for this to be a project that changes the future of housing in this country." (http://www.nytimes.com/2007/01/17/nyregion/17housing.html?ei=5090%26en=0f11db3693c7e52f%26ex=1326690000%26adxnnl=1%26partner=rssuserland%26emc=rss%26adxnnlx=1200086044-Ere2IvcaJMQIxxH3vhUz3A)

The project developer is Jonathan Rose Companies (www.rose-network.com) and co-developer Phipps Houses. The Jonathan Rose Companies have been a leader in green affordable housing in the U.S. for over a decade.

The Jonathan Rose Company was founded in 1989 by a third-generation New York real estate developer with a focus on sustainable, mixed-use communities in urban areas. Jonathan Rose shares a similar vision for green affordable housing as Enterprise Community Partners (www.enterprisecommunity.org). According to Paul Freitag, director of development for Jonathan Rose Companies, in a story by Nancy A. Ruhling in New York House Magazine (http://www.housemedianetwork.com/archive/article.php?issue=53&dept=92&id=870&pg=2), "We believe there should be no difference in cost to make affordable housing green" and "This is partly a function of the greater availability of green and energy-efficient products and also because in affordable housing we look to green responses that are not high-tech. We look for solutions that are integral to the architecture."

Most importantly, Freitag continues, "One of the most energy-efficient things you can do is to promote individual accountability in energy consumption" and "When residents become responsible for their own utility bills, their energy usage drops dramatically. Installing individual meters is a very green act, whether in condos or rental units." In addition, he correctly highlights that, "The affordable-housing population tends to include people impacted with health issues, particularly asthma, so we're interested in how green design can promote health and a healthy lifestyle."

I could not agree more with Jonathan Rose Company's assessment of the need for green affordable housing.

The "green aspects" of the development include the following:

- Rooftops that will be used to harvest rainwater, grow fruits and vegetables, and provide exercise and relaxation space for residents;

- The complex is oriented so that the height of the buildings steps up from south to north, bathing the entire site in sunlight. This design also became the origin for the series of green roofs that also recapture the landscape lost by constructing the buildings; and

- Cross ventilation, solar shades, non-toxic paints, rapidly renewable wood products, high-efficiency mechanical systems, energy-conserving appliances, and renewable energy strategies, including solar voltaic canopies.

As one might expect there is a strong stakeholder engagement aspect to the project. According to Freitag, "Community groups also are a valuable, and sometimes overlooked, resource and we develop a number of our projects in partnership with local community groups. Many of the opportunities for green development are in emerging neighborhoods, and through these partnerships our projects incorporate the concerns of the local community."

Finally, the Jonathan Rose Company views New York City as an ideal location as "It's already so far ahead in terms of population density and infrastructure, including mass transit," and, to one of the key points in this book, "One of the most green acts you can do is to build in the middle of a dense city grid. New York City is very innovative and open to green design, particularly in affordable housing." I could not agree more.

Another interesting and insightful aspect of the project is the funding sources for the project. Funding is from the following entities:

- Federal Low Income Housing Credits (LIHC)

- NYS Low Income Housing Credits (SLIHC)

- NYC Housing Development Corporation (HDC)

- Affordable Cooperative Program

- NYS Affordable Housing Corporation

- New Market Tax Credits (NMTC)

- Enterprise Green Community Grant

- NYSERDA Home Depot Grant.

Enterprise Community Partners

I have previously mentioned Enterprise Community Partners (Enterprise) (www.enterprisecommunity.org) and want to expand on their vision and projects. They are a fascinating organization with a dual not-for-profit and for-profit structure.

Enterprise had not been on the media radar until recently when Fast Company (December 2008) ran an article on them. What is revolutionary green about Enterprise is its focus on green affordable housing with many sites built on remediated brownfield sites.

Briefly, Enterprise has built a successful business by investing equity capital, pre-development lending, mortgage financing, and development grants to house low- and moderate-income families in the U.S. It helped create the low-income-housing tax credit that provides a means for businesses to profitably invest in affordable housing projects. This housing tax credit has driven the vast majority of affordable housing in the U.S.

One of Enterprise's most significant and successful programs has been Green Communities. The $555 million program is led by Enterprise (Brownfields Bulletin, November 14, 2006, "Building Green on Brownfields") and the National Resources Defense Council (NRDC) to build an estimated 8500 affordable homes that "promote health; conserve energy and other natural resources; and provide access to schools, jobs, and services." This program provides criteria outlining a standard for constructing green affordable housing, including the redevelopment of brownfields and the adaptive reuse of existing structures.

According to Dana Bourland, director of Green Communities for Enterprise, interviewed in Brownfields Bulletin, November 14, 2006, "Building Green on Brownfields," "Green affordable housing is achievable and evident in developments across the country."

Also according to Enterprise, the vision for Green Communities "is to fundamentally transform the way we think about, design, and build affordable homes." The focus of the Green Communities program is on the use of environmentally sustainable materials, reduction of negative environmental impacts, and increased energy efficiency.

Enterprise also promotes design principles and the use of materials that safeguard the health of residents and locations that provide easy access to services and public transportation.

The Green Communities Criteria (www.greencommunitiesonline.org/about/criteria), were developed by Enterprise in collaboration with:

- Natural Resources Defense Council
- American Institute of Architects
- American Planning Association
- National Center for Healthy Housing
- Southface Energy Institute
- Global Green USA
- Center for Maximum Potential Building Systems
- Experts associated with the U.S. Green Building Council

After the initial few years Enterprise has achieved the following:

- "$570 million in equity, loans, and grants
- 250 green affordable housing developments preserved or created
- 11,000 green affordable homes preserved or created
- 3000 housing professionals trained in green development
- 20 cities and states encouraged to adopt greener standards and policies"

There are numerous Enterprise projects that highlight their approach to transforming brownfield sites into green affordable housing. The Sara Connor Court project in Hayward, California is one example of a successful project.

The original site included a milk and juice processing plant, dry cleaner, and gas station with associated contaminated soils. These environmental impacts were remediated by the developer, Eden Housing.

The project consists of four buildings with a large open space and playground. The development consists of 59,570 square feet of residential space and 2175 square feet of community areas. The "green attributes" of the project include:

- Raised heel roof trusses offering both structural stability and room for more insulation;
- All appliances are Energy Star® rated;
- Water-efficient toilets, faucets, and showerheads;
- Natural linoleum flooring and low-emissions carpet;
- Low-VOC interior paints, adhesives, and sealants;
- Shade trees and California native or drought-tolerant plants used in landscaping;

- Playground surface made with recycled content manufactured from old tires; and
- Overall the project was built to be 20 percent more energy efficient than required by California's Title 24-2001 building energy standards.

The property includes such amenities as barbeque and picnic areas, a community room, a computer lab, and a landscaped courtyard with seating and play areas. The developer, Eden Housing, offers youth and adult programming to facilitate educational and economic advancement and promote a sense of community on the property. The comprehensive programs include after school and summer educational programs for children, financial literacy and homebuyer training, technology programming, and a resident scholarship program.

As with most projects, the financing was from several sources: City of Hayward, Redevelopment Agency of the City of Hayward; Lenders for Community Development; Silicon Valley Bank; Enterprise Tax Credits; Green Building in Alameda County Grant; Friendly Landscaping Grant; Enterprise Green Communities Grant; and a Home Depot Foundation Affordable Housing Built Responsibly.

CHAPTER
10

The Next 20 Years

INTRODUCTION

We are fortunate to live in "interesting times." I suspect that we are just warming up and the next 20 years should see dramatic changes in urbanization, the collision of energy demands with the "de carbonization" of industry, and perhaps the renaissance of "rust belt" U.S. cites and the reinvention of the suburbs.

If we are to continue urbanizing our society, as trends strongly suggest, it will be the brownfield sites that are redeveloped. The benefits of green brownfields development are numerous. Key benefits are: an established infrastructure and associated costs savings, lower environmental footprint than a suburban development (compare the carbon footprint per capita of New York with other U.S. cities—ranked 4th by the Brookings Institute based upon 2005 data. http://www.brookings.edu/reports/2008/~/mdia/Files/rc/reports/2008/05_carbon_footprint_sarzynski/carbonfootprint_report.pdf), and the social benefits of a community.

In the U.S. alone: there are an estimated 450,000 brownfields sites; 5 million acres may be abandoned industrial properties in urban areas, 20 to 50 percent of all industrial properties contain "environmental hazards," and for every acre of brownfileld sites re-developed approximately 4.5 acres of green space are preserved (EPA, National Brownfield Association, Environmental Law Institute, http://nreionline.com/mag/real_estate_brownfields_bloom).

The redevelopment of brownfield sites into "smart green urban growth" or "sustainable urban development" (a term used by the Northeast-Midwest Institute) goes well beyond green building practices.

A collection of "factoids" illustrating the full value of these "sustainable urban development" sites follows:

- Urban development avoids sprawl-related environmental impacts;

- When compared to suburban development patterns, compact infill redevelopment produces substantial air quality and energy-related benefits.

- Compact, mixed-use, interconnected, and pedestrian-friendly neighborhoods with transportation choices, a balance of homes, jobs, schools, and other uses can help reduce the need to drive and foster "walkability." Areas like these, whether dense city cores or small-town neighborhood streets on a grid, generate less vehicle travel (vehicle miles traveled—VMT) because people drive shorter distances and have to drive less due to the ability to walk or take transit.

- The average daily VMT for the 10 most sprawling metropolitan areas is 27 compared to 21 for the 10 most compact metropolitan areas.

- If 60 percent of new growth by 2050 is accommodated in "compact urban development," travel-related GHGs would be cut by 7 to 10 percent or 85 million metric tons of CO_2.

- Moving from the suburbs to an urban compact neighborhood is equivalent to driving a hybrid: hybrid fuel efficiency saves 2 metric tons of CO_2 relative to average vehicle fuel efficiency; urban compact neighborhoods save 2.1 metric tons of CO_2 via lower VMTs.

- According to a study commissioned by the American Council for an Energy Efficient Economy, shifting just 10 percent of new U.S. housing starts to smart growth would save 4.95 billion gallons of gasoline, 118 million barrels of oil, 59.5 mmt CO_2, and $220 billion in household expenses over 10 years.

- In terms of energy consumption, a "smart location" outperforms even the greenest sprawl house with hybrid cars. (136 million Btu/year versus 158 milllion Btu/year.)

- A report released by Urban Land Institute (ULI) documents that compact urban development, as an alternative to sprawl, could reduce VMT by 20 to 40 percent, or 30 percent as an average. This translates into a reduction of driving-related greenhouse gases by 7 to 12 percent by 2050.
- Another study by ULI reviewed the evidence of the relationship between density and VMTs—most studies reviewed indicate that, when controlling for factors such as income, any doubling of density corresponds to lowering of VMTs by about 25 to 30 percent.
- A comparison of the highly dense North Beach in San Francisco (100 households/residential acre) to low density suburban San Ramon (three households/residential acre) found that North Beach reduced VMTs by 75 percent.
- A Seattle study by Lawrence Frank found that households located in the most interconnected areas of Seattle generated less than half the driving of households located in the least-connected areas of the region, holding true after adjusting for household size, income, and vehicle ownership.
- A Center for Clean Air Policy study found that VMTs were an estimated 25 percent lower for an urban 20-unit per acre development than a suburban four-unit per acre development.
- An Atlanta regional study found that people who live in more walkable neighborhoods—with a mix of housing types and streets that connect to shops, offices, and other destinations— drive 30 percent less (accounting for 20 percent lower GHG emissions) than those in conventional auto-oriented settings, even when they own the same number of cars at the same rate.
- A King County, Washington, study concluded that urban "interconnected neighborhoods," defined by density, frequency of intersections, and grid street patterns, reduced VMTs by 26 percent relative to a suburban spread development model.
- Brownfield projects, as a subset of urban redevelopment activities, have demonstrated similar benefits in reduced VMTs. A study carried out by the U.S. Conference of Mayors compared development of brownfields and greenfields in Baltimore and Dallas. It concluded that brownfield redevelopment saved between 23 and 55 percent of VMTs.
- USEPA's study of Atlantic Station in Atlanta, a mixed-use redevelopment of the Atlantic Steel brownfield site, found VMT savings of a similar magnitude—14 to 52 percent.

- A study in the Bay Area by the Metropolitan Transportation Commission found that for people who both live and work within half a mile of a rail or ferry stop, 42 percent of them commute by transit. For those who neither work nor live within such proximity, the number falls to 4 percent. Elsewhere, individuals living in higher-density neighborhoods that include convenient access to transit, as well as pedestrian and bicycle-friendly features, reduce their driving by 15 to 50 percent.

- Urban redevelopment/brownfield projects generally use existing infrastructure and can be credited with energy savings associated with building and maintaining infrastructure when compared with greenfield development. A Center for Neighborhood Technology study found that the cost of providing infrastructure (roads, water, sewer, electricity, etc.) to a greenfield site averages $50,000 to $60,000 per unit, compared to $5,000 to $10,000 per unit for a brownfield or greyfield site. If energy use parallels costs, the comparative energy savings are substantial.

- Distributed energy provides energy savings attributable to use of distributed energy, combined heat and power (CHP), and/or other alternative or more efficient fuels. Because many cities have waste-to-energy plants that serve centralized areas, this is another source of lowered demands for fossil fuels and lowered emissions. One study concluded that one 1500-ton-per-day waste-to-energy facility in the northeast saved 270,000 tons of carbon-dioxide-equivalent emissions annually. However, it is not known how much of this savings is specific to serving urban core and brownfield areas.

- An indirect energy benefit of urban infill and brownfield redevelopment is the protection of "carbon sinks," i.e., greenfields that would have been developed absent the urban redevelopment project. No attempt has been made to quantify this factor, but it should be noted that tree-planting and reforestation are elements in some state strategies to reduce GHGs. Brownfield projects were analyzed for land utilization and density in a 2001 George Washington University study which found that, on average, one acre of brownfield redevelopment equaled 4.5 acres of greenfield development. Thus, it makes sense that urban redevelopment, because of the indirect benefits of saving greenfields, should be viewed as part of climate change plans.

THE NEXT 20 YEARS' TRENDS

So what are the big trends that will influence how these green brownfield sites are developed? In the next 20 years greening will no longer mean just reducing the environmental footprint of these sites. When these sites are redeveloped the questions that will be asked include: can they create green jobs, can they be built "carbon neutral," and will they provide affordable green housing?

I believe we will see the following big trends drive an increasingly greener brownfield redevelopment. The big trends that will influence how green these brownfield sites become and what the next couple of decades look like are diverse yet reinforcing.

- Re-industrialization—Innovation and creative destruction
- Green affordable housing and green jobs
- Revolutionary building materials
- Climate change and carbon neutral cities
- Reinventing U.S. cities
- China—The urbanization of China
- Connected cities
- Renewable energy

The greening of brownfield sites will create greater value than just remediating properties. The growing consensus is that if brownfield sites are not built green, "value is left on the table."

Re-Industrialization—Innovation and Creative Destruction

Let's start with a "30,000 ft view" of what is happening to industry. I believe we are in the midst of a wave of innovation and the creative destruction of industries with sustainability as a major driver in this transformation.

I believe innovation and creative destruction is the real macro driver in changes in our global economy and industrialization. Declining industrial sectors are resulting in the closing of industrial sites and shifting urban centers, more brownfield sites.

This is essentially the "re-industrialization" or the "second industrial revolution" with resultant fallout in the shuttering of businesses and industries and the creation of new industries.

The expression "creative destruction" was first introduced by Joseph Schumpeter in his book "Capitalism, Socialism, and Democracy" in 1942. Schumpeter argued that innovation and technological changes come from entrepreneurs. In addition he asserted that big companies are the ones that drive innovation as they have the resources and capital to invest in research and development.

Today we see both large companies and "startups" driving innovation. One only has to look at the U.S. automobile sector and automobile startups in Silicon Valley (Tesla, Fisker Motors for example) to get a sense of how the innovation is unfolding.

Stu L. Hart and Mark B. Millstein in their 1999 article titled, "Global Sustainability and the Creative Destruction of Industries" (Sloan Management Review, ISSN 0019-848X, Vol. 41, No. 1, 1999) linked Schumpeter's business concept of creative destruction to the business drivers of sustainability. I agree with Hart and Millstein.

Many people think that the current "destruction" we're seeing is exclusively the result of the financial meltdown of 2008 and 2009, but I believe creative destruction was in play well before the financial meltdown. The meltdown is accelerating this destruction, but innovation is alive and well, and those who can survive this unprecedented economic environment will thrive.

Hart and Milstein argue that "The emerging challenge of global sustainability will catalyze a new round of 'creative destruction' that innovators and entrepreneurs will view as one of the biggest business opportunities in the history of commerce."

Creative destruction is having a significant impact on industries and cities—not just increased idle industrial sites but a realignment of urban centers as old industries struggle.

Green Affordable Housing and Green Jobs

The old business paradigm was that resources were limitless and stakeholders were of no concern. The new paradigm of sustainability recognizes that resources are limited and that stakeholders play a crucial role in building a sustainable business. Stakeholders have an enormous impact on a company's license to operate. Engaged and supporting stakeholders will provide support for a project and the development partners.

Stakeholders include community and national NGOs with a mission to develop green affordable housing. Increasingly, NGOs focused on green affordable housing are part of an effort to transform brownfield sites into green development projects which also create green jobs.

On a large scale Sustainable South Bronx has been a driving force in creating a vision for the redevelopment of the South Bronx creating green jobs along the way.

Organizations such as Enterprise Community Partners (Enterprise) and its Green Communities Program are fundamentally transforming how green affordable homes are designed and built. The Enterprise program was the first U.S. national green building program developed for affordable housing. The focus of the program is on the use of "environmentally sustainable materials, reduction of negative environmental impacts, and increased energy efficiency."

Enterprise is focused on creating communities that utilize green designs and materials that safeguard the health of residents, and locations that provide easy access to services and public transportation are emphasized. The program is also designed to assist developers, investors, builders, and residents in making the transition to green affordable housing.

The Green Communities program was created with input from several stakeholders including U.S. experts in environmental issues, public health, and green building practices. Enterprise provides grants, loans, tax credit equity, and training/technical assistance in support of creating green affordable housing.

Enterprise established its "Green Communities" criteria (http://www.greencommunitiesonline.org/about/criteria). These guidelines were developed with input from the following stakeholders:

- Natural Resources Defense Council
- American Institute of Architects
- American Planning Association
- National Center for Healthy Housing
- Southface Energy Institute
- Global Green USA
- Center for Maximum Potential Building Systems
- Experts associated with the U.S. Green Building Council

The Enterprise Green Communities program has been successful beyond initial expectations. Some of the statistics (according to Enterprise) include:

- $570 million in equity, loans, and grants
- 250 green affordable housing developments preserved or created
- 11,000 green affordable homes preserved or created
- 3000 housing professionals trained in green development
- 20 cities and states encouraged to adopt greener standards and policies

These early results have shown that affordable homes and developments can attain a high standard of environmental sustainability for a modest additional cost through the Green Communities approach. Their model delivers a comprehensive package of environmental resources to developers and policymakers that can be adapted to local conditions, combined with other private and public resources, and used to achieve transformational change in local markets across the country.

Enterprise has also taken the Green Communities program further by creating a "Green Communities Offset Fund" which supports the offsetting of carbon emissions (more on carbon neutral cities later in this section).

The Enterprise Green Communities Offset Fund works as follows (from Enterprise):

- "Enterprise raises charitable contributions to support the development and rehabilitation of green affordable homes that generate lower carbon emissions.
- These contributions are used to purchase carbon offsets that are measured and verified from affordable housing sponsors.
- Documentation is provided to demonstrate to fund contributors that their funding has helped create green affordable homes and reduce carbon emissions."

I believe that going forward "green brownfield sites" will see an increasing emphasis on green affordable housing. Low income families are those that can benefit the most from a lower cost of operating their homes. Moreover, green affordable housing links very well to the creation of green jobs, my next topic.

Green Jobs

You wonder, green affordable housing and green jobs as part of the greening of brownfield sites? The revitalization of abandoned urban sites and the associated opportunity to develop green affordable housing and create green jobs are directly linked.

For example, the U.S. Brownfields Economic Development Initiative (BEDI) created by the U.S. Department of Housing and Urban Development (http://www.hud.gov/offices/cpd/economicdevelopment/programs/bedi/) provides grant programs to promote economic and community development. The BEDI program assists cities in the redevelopment of contaminated (real or potential) abandoned urban industrial and commercial properties.

The BEDI program creates the linkage between the redevelopment of abandoned urban brownfield sites and the creation of economic development in the community. The focus is on creating economic opportunities for low- and moderate-income sectors of the community and the creation or retention of businesses and jobs and a resultant increase in the local tax base.

BEDI funds are used to finance projects and activities that will provide near-term results and clear economic benefits. The program is not designed for the site acquisition and/or remediation where there is no immediate plan of redevelopment. The real goal of the program is to aggressively promote the return of brownfields to productive economic use through financial assistance to public entities in the redevelopment of brownfields, and enhance the security or improve the viability of a project financed.

The BEDI program is designed to be used with Section 108, the loan guarantee provision of the Community Development Block Grant (CDBG) program. According to HUD, "BEDI projects must increase economic opportunity for persons of low- and moderate-income or stimulate and retain businesses and jobs that lead to economic revitalization."

A "revolutionary" legislative development in the U.S. was the U.S. Green Jobs Act (HR 2847) of 2007 (http://www.govtrack.us/congress/bill.xpd?tab=summary&bill=h110-2847) that authorized $125 million per year to create "Energy Efficiency and Renewable Energy Worker Training Program as an amendment to the Workforce Investment Act (WIA)." The Green Jobs Act became Title X of the Energy Independence and Security Act (referred to as the "2007 Energy Bill"), which Congress passed and the president signed in late 2007. The program will be administered by the U.S. Department of Labor (DOL) in consultation with the Department of Energy.

The Green Jobs Act (GJA) is designed to identify skills needed, develop training programs, and train workers for jobs in a range of industries as we "green the economy." These new green jobs are part of a movement to build and retrofit energy efficient buildings, develop renewable energy, build energy efficient vehicles, biofuels, and manufacturing that "produces sustainable products and uses sustainable processes and materials." The best way to put this is that the program will focus on creating "green pathways out of poverty."

Urban redevelopment of brownfield sites and "green pathways" out of poverty are directly linked goals. Urban communities may have the chance to turn urban blight into green affordable housing and a place for green jobs.

As part of the relatively recent push to develop green industries there will be a demand for skilled and unskilled workers. The GJA is focused on meeting the demand for new jobs in emerging industries and related services such as retrofitting buildings, installing solar panels, maintaining wind farms, manufacturing component parts, building new facilities and infrastructure, etc. The GJA will identify the skills and provide training for these workers (living in green affordable housing perhaps).

The GJA authorizes (although not actually funded as yet) spending for the following five green job programs:

- National Research Program—The Department of Labor (DOL), acting through the Bureau of Labor Statistics, will collect and analyze the labor market data necessary to track workforce trends and identify the types of skills and green jobs needed.

- National Energy Training Partnership Grants—The DOL will award competitive grants to nonprofit partnerships to carry out training that leads to economic self-sufficiency and to develop an energy efficiency and renewable energy industries workforce.

- State Labor Market Research, Information, and Labor Exchange Research Program—The DOL will award competitive grants to states to administer labor market and labor exchange information programs, in coordination with the "one-stop delivery system." Activities will also include the identification of job openings; the administration of skill and aptitude testing; and counseling, case management, and job referrals.

- State Energy Training Partnership Program—The DOL will award competitive grants to states to enable them to administer, via the state agency that administers their employment service and unemployment insurance programs, renewable energy and energy efficiency workforce development programs.

- Pathways Out of Poverty Demonstration Program—The DOL will award competitive grants to training partnerships that serve individuals under 200 percent of the federal poverty line or a locally defined self-sufficiency standard. The partnerships will include community-based organizations, educational institutions, industry, and labor; demonstrate experience implementing training programs and recruit and support participants to the successful completion of training; and coordinate activities with the WIA system.

Revolutionary Building Materials

In our quest to "redesign everything" we are seeing the birth of new materials for construction. The building and construction sector is witnessing the emergence of companies dedicated to innovative processes and materials and well established companies driving innovation to improve product performance (increased energy efficiency, increased durability, and reduced carbon emissions). These new materials will shape how we build green.

This overview is not meant to be exhaustive. New companies and new products are emerging rapidly. Instead it is meant to provide a feel for the innovation in the building and construction materials sector. Moreover, new companies are creating new jobs, in fact green jobs.

Serious Materials

Let's begin with Serious Materials (www.seriousmaterials.com) as an innovative manufacturer of energy efficient building products and the creator of green jobs.

According to Serious Materials "The world today produces over 30 billion tons of greenhouse gas emissions per year." The company, through its products, "aims to save 1 billion tons of greenhouse gas emissions every year."

The portfolio of Serious Materials products is varied but all designed to increase energy efficiency (and as a result reduce greenhouse gas emissions through new manufacturing processes and actual use). The value proposition from Serious Materials is that by re-engineering building and construction products they can reduce energy use and greenhouse gas emissions while creating "thousands of green collar jobs."

Their product portfolio is as follows.

EcoRock™. The production of standard gypsum drywall (a process dating back to 1917) results in up to 20 billion pounds of CO_2 emissions per year. Drywall is the third-worst greenhouse gas producer in building materials (www.seriousmaterials.com).

The EcoRock™ product is non-gypsum, "green alternative" to standard drywall that performs and is used just like drywall. The product requires approximately 80 percent less energy to manufacture with a resultant reduction in greenhouse gas emissions.

This reduction is achieved by not using heaters or dryers in production, nor calcining processes. EcoRock also uses 85 percent post-industrial recycled content and is fully recyclable. A plant built with EcoRock reduces energy use to "a minimum."

According to Serious Materials, "Over 30 billion square feet of drywall are produced each year in the U.S. and Canada. A single sheet uses between 100,000 and 400,000 Btus of energy to produce, depending on the age of the plant, producing 16 pounds of greenhouse gases per sheet. Approximately one-quarter of the cost of a gypsum drywall panel is tied to energy, with likely long-term cost increases; carbon taxes are common in Europe and may come to the U.S."

ThermaProof Energy-Saving Windows. Windows represent the single largest opportunity for improvement in the built environment with approximately 39 percent of all emissions tied to building operations, with 38 percent of that for heating and cooling. Up to 40 percent of that energy is wasted through energy loss. The energy loss is calculated by Serious Materials at over 250 million tons of greenhouse gas emissions per year.

The ThermaProof windows are rated from R-5 to R-11 which is up to four times higher than major brands, and USEPA Energy Star requirements.

QuietRock®. The QuietRock® soundproof drywall will "reduce material use, enhance livability and support denser, more sustainable urban construction." The Serious Materials QuietHome® Windows also reduce outside noise.

So where are the "green jobs?" Serious Materials currently has manufacturing and R&D operations in: Sunnyvale, California; Newark, California; Boulder, Colorado; Vandergrift, Pennsylvania and most recently in Chicago, Illinois (http://www.seriousmaterials.com/html/vice-president-biden-visits-serious-materials.html).

BASF

In contrast to Serious Materials, BASF (www.basf.com) is the world's largest chemical company with a memorable tag line, "We don't make a lot of the products you buy. We make a lot of the products you buy better.®"

BASF products range from oil and gas to chemicals, plastics, performance products, agricultural products, and fine chemicals. BASF works closely with its customers to create "high value products and intelligent solutions." These intelligent solutions include building and construction products that are energy efficient (with resultant lower greenhouse gas emissions) and durable. In other words, "sustainable construction."

I highlight BASF because we are seeing a move to green building products from both new innovative companies and well established ones like BASF (full disclosure that BASF is a client). I will touch on the BASF Near Zero Energy Home (NZEH) and their "sustainable construction" products.

In April 2006, BASF opened the NZEH in Paterson, New Jersey. The home was designed to be energy efficient, affordable, and durable. According to BASF the home is 80 percent more efficient than a conventionally built home. BASF took the lead in building the home but partnered with about 150 customers and "strategic allies."

The project was designed to achieve a 95.5 HERS ENERGY STAR® score, an "unprecedented accomplishment" in New Jersey ENERGY STAR history. The home was ultimately donated to St. Michael's Housing Corporation and in turn to a local family with a quadriplegic boy to occupy (the project also highlights accessible design).

The home was part of the BASF Better Home, Better Planet Initiative which leverages the experience gained at the NZEH and "sustainable construction" products. The BASF "sustainable construction" initiative includes more than 600 technologies, product offerings, and 75 construction solutions (www.basf.com).

Sustainable construction consists of:

- "Building envelope technologies for roofs, walls and foundations;
- HVAC & plumbing components for high-efficiency comfort conditioning and plumbing systems;
- Renewable energy systems;
- Concrete, additives & infrastructure technologies for improved durability, strength, and lifecycle cost; and
- Coatings to protect against weather, water, chemicals, and wear."

Calera Corporation

Now let's look at radical materials by examining a company by the name of Calera.

Calera Corporation (www.calera.biz) is a venture capital funded company with a focus on the "the green-age built environment" and "sustainable, renewable new building materials with ancient origins."

The company basically developed a nature-mimicking (a process that converts CO_2, such as from a coal power plant into cement). The process removes CO_2 from the atmosphere in the process of making cement in a conventional manner (high CO_2 intensity process). The Calera cement can be mixed with conventional Portland cement.

The founder of Calera, Brent Constantz, has stated that "For every ton of cement we make, we are sequestering half a ton of CO_2. We probably have the best carbon capture and storage technique there is by a long shot."

The Calera process mimics marine cement, which is produced by coral when making shells and reefs. The calcium and magnesium in seawater is used to form carbonates at normal temperatures and pressures.

The company is really making chalk and the Calera cement can be used as a replacement for traditional Portland cement that is usually blended with other materials to create construction concrete.

Imagine building materials that sequester CO_2 during the manufacturing process.

Other R&D groups are working on solving the same with energy and carbon intensive construction materials such as cement. In

January 2007 the MIT Civil and Environmental Engineering Department announced it was using nanotechnology to reduce CO_2 emissions in the manufacturing process (the production of cement, the primary component of concrete, accounts for 5 to 10 percent of the world's total carbon dioxide emissions). Interestingly enough, the research was partially funded by the Lafarge Group.

Climate Change and Carbon Neutral Cities

In the U.S., residential and commercial buildings alone account for 39 percent of the carbon emissions in the United States. Transportation accounts for one-third of U.S. emissions and industry is responsible for 28 percent. An effective climate strategy must focus on reducing carbon emissions from all three sectors.

There is an increasing recognition that buildings and cities are central to how we approach reducing carbon emissions. Green building initiatives are part of the solution but taking a holistic approach to reducing carbon emissions at the city scale is where we are headed.

In the U.S. we are quickly moving toward federal regulation of carbon either through the USEPA or through congressional action. While it is unlikely these regulations will impact buildings or cities, carbon and climate awareness has increased. As a result, the public (and private sector) is thinking more about climate change and carbon emissions.

A 2008 study by the Brookings Institute titled, "Shrinking the Carbon Footprint of Metropolitan America" by Marilyn A. Brown, Frank Southworth, and Andrea Sarzynski (as part of the Brookings Institute Metropolitan Policy Program) outlines a federal approach to reducing the carbon footprint of U.S. cities.

Their thesis is that with an ever growing population and overall expanding economy (not by late 2008 or 2009 however), America's "settlement area" is increasing and as it does there is an increase in travel (primarily automobile), increased energy use and a resultant increase in carbon emissions.

Therefore, where we work, live and "recreate" are important to our overall sustainability and energy policies. The authors propose the following:

- "Promote more transportation choices to expand transit and compact development options;

- Introduce more energy-efficient freight operations with regional freight planning;

- Require home energy cost disclosure when selling and "on-bill" financing to stimulate and scale up energy-efficient retrofitting of residential housing;

- Use federal housing policy to create incentives for energy- and location efficient decisions; and

- Issue a metropolitan challenge to develop innovative solutions that integrate multiple policy areas."

Progress in addressing climate change and carbon emissions is being made by cities on an international level. One important organization focused on addressing sustainability and climate change at the city scale is the "International Council for Local Environmental Initiatives" or ICLEI. ICLEI was founded in 1990 when over 200 local governments from 43 countries participated in the "World Congress of Local Governments for a Sustainable Future." ICLEI renamed itself and refocused its mission to "ICLEI-Local Governments for Sustainability." The ICLEI mandate is now to address broader sustainability issues for global cities.

ICLEI runs successful and widely recognized regional and national programs in Australia, Canada, Europe, Japan, Latin America, Mexico, New Zealand, South Africa, South Asia, Southeast Asia, and the United States.

The ICLEI "Cities for Climate Protection™ (CCP) campaign by ICLEI is designed to support the development of policies and metrics to address reducing greenhouse gas emissions and promote overall urban sustainability. The CCP program has widespread support with over 700 local governments participating and integrating climate change mitigation and greenhouse gas reductions into decision making processes and policies. ICLEI estimates that CCP participants account for approximately 15 percent of global greenhouse gas emissions.

As with most international programs to address climate change the participants begin with a commitment to reduce greenhouse gas emissions within their jurisdictions. ICLEI then provides support for these government entities to reduce greenhouse gas emissions by moving toward achieving five milestones. ICLEI also provides software tools to assist cities in goal setting and measuring performance.

The five milestones are (www.iclei.org):

- "Milestone 1. Conduct a baseline emissions inventory and forecast. Based on energy consumption and waste generation, the city calculates greenhouse gas emissions for a base year (e.g., 2000) and for a forecast year (e.g., 2015). The inventory and forecast provide a benchmark against which the city can measure progress.

- Milestone 2. Adopt an emissions reduction target for the forecast year. The city establishes an emission reduction target for the city. The target both fosters political will and creates a framework to guide the planning and implementation of measures.

- Milestone 3. Develop a local action plan. Through a multi-stakeholder process, the city develops a local action plan that describes the policies and measures that the local government will take to reduce greenhouse gas emissions and achieve its emissions reduction target. Most plans include a timeline, a description of financing mechanisms, and an assignment of responsibility to departments and staff. In addition to direct greenhouse gas reduction measures, most plans also incorporate public awareness and education efforts.

- Milestone 4. Implement policies and measures. The city implements the policies and measures contained in its local action plan. Typical policies and measures implemented by CCP participants include energy efficiency improvements to municipal buildings and water treatment facilities, streetlight retrofits, public transit improvements, installation of renewable power applications, and methane recovery from waste management.

- Milestone 5. Monitor and verify results. Monitoring and verifying progress on the implementation of measures to reduce or avoid greenhouse gas emissions is an ongoing process. Monitoring begins once measures are implemented and continues for the life of the measures, providing important feedback that can be used to improve the measures over time."

According to ICLEI these milestones provide a flexible framework for cities to adopt which has promoted its acceptance on a global level.

While ICLEI is working with global cities to reduce greenhouse gas emissions it has also recognized (and I concur) that we are likely too late to mitigate the impacts of greenhouse gas emissions on climate. As a result it is prudent risk management to plan on adapting to climate change. Changes in weather patterns and rising sea levels will impact buildings and infrastructure.

In recognition of this need, ICLEI developed a strategic plan to address climate change adaptation in 2006. ICLEI essentially recognized that despite the global recognition of the seriousness of climate change there was a "gap" between the talk and action.

Policy and decision-making that addresses climate change impacts must respond to the inherent complexity and uncertainty of this issue, and requires a high level of sophistication from risk management approaches and processes.

ICLEI has been very active in addressing adaptation and climate change. These include the following global initiatives:

- CCP Australia Adaptation Initiative
- CCP Europe Adaptation Initiative
- ICLEI USA Climate Resilient Communities™
- Pilot Project on Climate Adaptation in Canadian Municipalities
- Adaptation Working Group (to join email ccp@iclei.org)
- Representing local governments as observers to the United Nations Framework Convention on Climate Change

ICLEI is worth tracking as they have a bias for action and are addressing the big issues with regards to sustainability and climate change.

Let's take a look at cutting edge "carbon neutral" cities and the implications of such development projects.

Currently, the most ambitious is Masdar City in Abu Dhabi. A recent article in Technology Review Magazine (Technology Review) on April 16, 2009 provides insight into the rationale and goals of the project.

So, why would an "oil rich" country like Abu Dhabi build a zero emissions city? In a word, "vision."

The recognition that globally we are moving toward a carbon constrained world and that at some point oil will not be the dominant fuel is driving this experiment. The objective is to create a car free, carbon dioxide free and zero waste city. This is remarkable as it takes the concepts of OPL (One Planet Living) much further. This $15 billion investment is scheduled to be completed in 2016. About $4 billion is dedicated to the city's infrastructure and the balance will be funded from outside investors. The project is an investment by the government of Abu Dhabi and is part of the "Masdar Initiative." Planners predict that the development will attract 1500 clean-tech businesses, ranging from large international corporations to startups, and—eventually—some 50,000 residents.

A bit more on the rationale for this project, the Masdar Initiative is part of an ambitious plan to transform a resource-based economy (third-largest exporter of oil in the world) into one based on knowledge and expertise. Remember the discussion on innovation and creative destruction earlier in this chapter? This is innovation on a grand scale linking public policy and private initiatives.

Masdar in Arabic means "source" which is fitting as the overall public policy goal is to turn Abu Dhabi into the "Silicon Valley of alternative energy." The business model for Masdar City is to create a tax-free zone to attract clean-technology companies. The first tenant will be General Electric who plans to build a 4000-square-meter facility. In addition, the Masdar Institute is conceived as the Stanford University of Masdar and the curriculum is being developed with input from MIT in the U.S.

The scale of the city is impressive as it will cover approximately 6 square kilometers.

The key elements of the project include the following:

- Waste—Waste will be recycled or used for compost and sewage will be processed into fuel. Sewage will be treated and where possible processed into a dry renewable fuel for generating electricity.

- Water—Water will be recycled to save on energy costs from desalination.

- Transportation—The transportation system will include a light-rail line linking the development to downtown Abu Dhabi and the airport and a very innovative personal rapid-transit (PRT) system within the city. The PRT is a system of automated electric vehicles that will provide access to the light rail or car parking garages outside the city. The city will be about 7 meters above ground to accommodate the PRT. The PRT is designed to use less energy than conventional mass transit. The PRT consists of pods; an individual or a small group boards a pod and selects a destination. The pod then proceeds automatically to the destination. The pod follows a track which is connected to stations by on-ramps and off-ramps and a computer controls how the pods enter and exit the stations.

- Carbon—Zero emissions is the hallmark of the city and essentially drives most of the other environmental initiatives (transportation, waste, building design, etc.). The city will need to rely in part on fossil fuels (natural gas) when the solar panels can't produce electricity. The city is really a "zero net carbon dioxide emission" city. This means that the city will rely on the use of carbon credits to achieve carbon neutral status.

- Buildings—The buildings will have thick walls (30 cm with skins of copper foil to reflect light away from the buildings) to absorb heat during the day and will be densely packed closely together to provide shade for each other. Narrow passages between the buildings cool the buildings. Water consumption sill be reduced by 75 percent through recycling, the use of low-flow fixtures, and waterless urinals.

- Data management and monitoring—Data gathering and evaluation will be essential to measure both energy consumption and energy production. In addition, water use will be measured to the individual fixture. Also being considered is the use of RFID tags in security badges to gather information on the way people use water and energy.

This is more than an experiment. The builders and those funding the project want this to be a profitable business as they should. According to the project's director of property development, the project should be a sunk cost and if it is not profitable it is not sustainable (the key aspect of the "triple bottom line"). Most importantly, the project experience can be leveraged and translated to other development projects if it is profitable.

This may be in part a blueprint for other carbon neutral cities or retrofitting existing cities.

It is not just Masdar that is thinking about climate change and carbon. Paris is engaged in a design competition to envision a "post-Kyoto city." While not focused on being carbon neutral the design competition does drive thinking beyond mere infrastructure upgrades.

Recently Paris concluded a nine-month study on the future of metropolitan Paris. The study, which included several prominent architects, is the first phase in a plan to create a more sustainable, socially integrated model of "the post-Kyoto city." Ten scenarios were created (http://www.bustler.net/index.php/article/ten_scenarios_for_grand_paris_metropolis_now_up_for_public_debate). This visionary plan was set in motion in September 2007 and is titled ""La Cité de l'architecture & du patrimoine," the visionary plan for a new "Grand Paris."

The actual design process was initiated in 2008 and is being managed by a steering committee made up of representatives of the City of Paris, the Île-de-France Regional Authority and the Île-de-France Mayoral Association. In addition, the steering committed has the support of a scientific committee of 23 led by architect Paul Chemetov and geographer Michel Lussault. The French minister of Culture and Communication is responsible for the consultation process.

I find this project extremely exciting and visionary. The mere concept of a "post-Kyoto city" begins to move our thinking toward how to build (or rebuild) cities in a carbon constrained world. Perhaps this is just a dream but it has put "carbon constrained" on the minds of planners and architects on a grand scale.

The 10 plans are bold and visionary at a time when we need big thinking (what better time to rethink our future when in the depths of a global economic meltdown). Although there is no real sense of how this will be funded, the "exercise" does provide credible and in depth analysis of the uniqueness of Paris and its diversity (and challenges). One of the major challenges for Paris is the separation between the picture postcard Paris lying within the elevated freeway circling the city and the surrounding suburbs. The suburbs consist of blocks of apartments essentially cut off physically and socially from the city center (http://www.nytimes.com/2009/03/17/arts/design/17paris.html?_r=2&scp=1&sq=paris&st=cse).

The two major goals of the competition were to: create a plan for a greener, more sustainable city, and to break down the isolation between the outlying neighborhoods and the historic center. What I find very progressive about the project is the goal to address both environmental and social aspects of sustainability. We typically tend to just address the environmental issues as they are most easily visible and quantified.

I really can't stress this enough—there is no separation between environmental and social performance. Sarkozy has it right to address both in this visioning exercise. Regardless of how this project plays out, highlighting the issues and exploring solutions has value.

The plans are bold. Some of the ideas are:

- Mr. de Portzamparc. Demolish the Gare du Nord and the Gare de l'Est and replace them with a single massive train station just outside the city center. The station would link to London and Brussels and a new elevated maglev train that would run above the périphérique. It would also be the anchor a towering new global business district, a rival to La Défense.
- Sir Rogers. He proposes burying the railroad tracks that connect the Paris rail stations to the city, all underground. A system of public parks would cover these new underground tracks, connecting poor and middle-class neighborhoods. A new Métro line would ring the outer city and more trains would tie the system back to the historic center.
- Jean Nouvel. He proposes creating a green belt that would circle the entire city. "All future construction would be concentrated inside this belt, adding density to what are now sprawling, isolated communities. New towers would punctuate some of the outlying boulevards, adding visual markers where there are none. The outer ring would become a sort of 620-mile-long community garden, with residents tending their plots along an endless string of parks and fields. The idea is to give a powerful identity to the most anonymous parts of the city."
- Djamel Klouche. He proposes transforming the space underneath the Louvre pyramid into a bustling Métro hub, "making one of Paris's greatest cultural monuments the main entry point to the city center for its immigrant masses."
- Roland Castro. He suggests moving existing monuments, including the Élysée Palace, to the city's grittiest outlying neighborhoods.

Highlights of the plans are illustrated in Figures 10.1 through 10.9 (http://www.bustler.net/index.php/article/ten_scenarios_for_grand _paris_metropolis_now_up_for_public_debate).

Reinventing U.S. Cities

I recently read an article in the New York Times by Nicolai Ouroussoff (March 29, 2009) titled "Reinventing America's Cities: The Time is Now." The article makes a strong case for investing in the redevelopment of U.S. cites and ties into my early comment in this chapter regarding innovation and creative destruction (sustainability is driving innovation and the creative destruction of industries, realignment of urban areas, etc.).

Ouroussoff makes the case that the U.S. needs a bold vision for remaking decaying U.S. urban centers and highlights a few examples. The decay was the result of years of poor policies that promoted the development of sprawling suburbs at the expense of thriving urban centers. This decay has resulted in the creation of unsustainable cites at best and ghost towns at worst with gated communities and divisions between racial, ethnic, and economic classes.

He argues that we need a grand plan on the scale of the U.S. "Works Projects Administration" (WPR) established by Franklin D. Roosevelt in the 1930s or Dwight D. Eisenhower's 1956 National

Figure 10.1 Sir Richard Rogers, Rogers Stirk Harbour & Partners/London School of Economics/Arup team. Image: Rogers Stirk Harbour Partners.

Figure 10.2   Yves Lion, Groupe Descartes team. Image: © A. Grondeau.

Figure 10.3   Djamel Klouche, AUC team. Image: L'AUC.

Figure 10.4 Christian de Portzamparc, Atelier Christian de Portzamparc team. Image: Atelier Christian de Portzamparc.

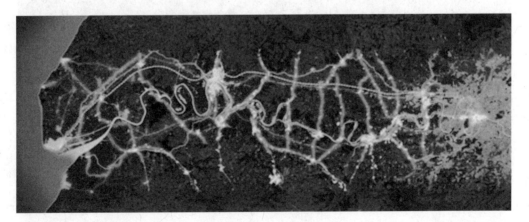

Figure 10.5 Antoine Grumbach, Agence Grumbach and Associates team. Image: Antoine Grumbach.

Figure 10.6 Jean Nouvel, representative of the Ateliers Jean Nouvel/Michel Cantal-Dupart/ Jean-Marie Duthilleul team. Image: Ateliers Jean Nouvel.

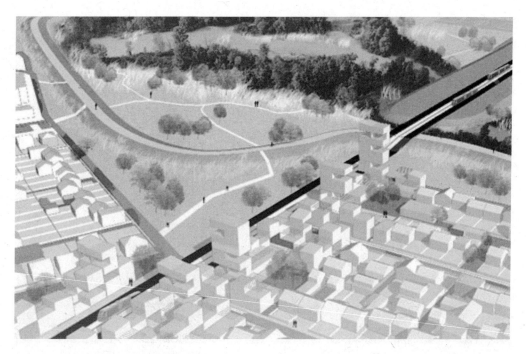

Figure 10.7 Bernardo Secchi and Poal Vigano, Studio 09 team. Image: Studio09 Bernardo Secchi Paola Viganò.

Figure 10.8 Finn Geipel, LIN team. Image: Linn Finn Geipel-Giulia/Andi.

Figure 10.9   Roland Castro, Ateliers Castro/Denissof/Casi team. Image: Atelier Castro Denissof Casi.

Figure 10.10   Winy Maas, MVRDV team. Image: MVRDV Grand Paris.

Interstate and Defense Highways Act. I agree. Rebuilding U.S. urban centers is in dire need of attention and worth the investment. It all ties together, innovation and new industries, redevelopment of brownfield sites, green affordable housing, and green jobs.

While these previous massive U.S. infrastructure projects had enormous economic benefits to the U.S., several "unintended consequences" emerged. The flight to the suburbs and the creation of suburban sprawl most notably, and the dissection of urban centers by highways stand out.

Ouroussoff examines the "ills" of four U.S. cities and envisions what redevelopment might look like.

New Orleans

The bottom line for New Orleans is that in many ways Hurricane Katrina provided urban planners and architects with the chance to re-examine the fabric of the city and think about how environmental and social issues (sustainability) could reshape the city going forward. The Urban Land Institute (ULI) proposed returning part of the city's low lying areas to natural conditions (wetlands) and moving some of the rebuilt housing to higher ground. Local architects began an effort to preserve some of the remaining local history including shotgun houses and the French Quarter. Local developers and world class architects have all been working on integrating sustainability into the redevelopment of the city.

So far not much real progress has been made as the interests of private developers, communities, local government, and national NGOs and consultants "battle" as to what is best for New Orleans.

The outcome is uncertain at this point. What is clear is that all stakeholders must align with a common purpose and vision for the city.

Los Angeles

Yes, Los Angeles is a city with enormous potential to embed sustainability into redevelopment. Recently the Los Angeles City Council revived an old vision from the 1930s by Frederick Law Olmsted which envisioned transforming the LA concrete river beds (about 51 miles) into a series of interconnected parks. A great plan, however with little to no funding. Currently there is about $6 million in state grants (bear in mind the current financial problems of California in 2009). This is coupled with some funding for the U.S.

Army Corps of Engineers for a feasibility study. A major portion of the project could be completed near downtown for about $100 million. Not a lot of money considering what the benefits would be to the residents of LA.

According to Ouroussoff the Wilshire Boulevard area is also viewed as a prime spot for redevelopment and a new approach. The recommendations for this area have focused on increasing mass transit and concentrating cultural institutions along Wilshire.

So it appears that the key opportunities in LA are to revitalize Wilshire and the LA River with increased open space and mass transit/cultural institutions.

Bronx, New York City

I highlighted the Bronx previously and believe it represents a great opportunity to highlight sustainable development and green job creation. Key issues on the Bronx are the Sheridan and Bruckner Expressways. Again according to Ouroussoff a plan was proposed in 1997 to expand the expressway's entry ramps which prompted a counterplan to dismantle the expressway. The removal of the expressway would open up 28 acres of land, extend local streets, add a riverfront park, provide up to 1200 units of affordable housing, create a new sewage facility, and restore wetlands along the river. Another great vision and plan but without the funding needed to make it happen.

Buffalo

Buffalo is a classic example of a U.S. "rust belt" city. The city is the location of great American architecture including parks designed by Olmsted.

The departure of the industrial base from Buffalo has led to urban flight and the demolition of parks and essential public spaces. This trend seems to be accelerating and not slowing down. Plans are being developed to expand the Peace Bridge to Canada which would demolish a working class neighborhood and unique local architecture. Currently there are no plans (and funding) to counter the plan. It remains to be seen if local preservationists can save this part of the city.

Despite the bleak picture for these urban areas Ouroussoff highlights some hope. He cites a recent announcement that the U.S. Departments of Transportation and Housing and Urban Development have created an urban task force to "promote the development of sustainable communities linked to public transportation." He also highlights the benefits of the National Infrastructure Bank which is a concept similar to the World Bank. The NIB would finance large-scale projects in the U.S. such as subways, airports and harbor improvements, hopefully with a sustainability focus. Unfortunately it is not clear if this will come to pass.

It is important to highlight a bright spot in our discussion of urban planning and vision. The PLANYC 2030 for New York provides hope that a city can actually transform itself into a more sustainable place to live and work (www.nyc.gov). New York is not the only city with bold plans to create a more sustainable future but it is one of the most comprehensive.

Mayor Bloomberg announced a plan titled PLANYC 2030 with a view that New York will have another million residents by 2030 and the city needed to start planning now. These million new residents are projected to need about 265,000 new housing units. The key element in the plan was what infrastructure will be needed in a city where decay is prevalent.

The city will address:

- The long-term environmental and health effects of growth (both past and present);
- How the city's energy and transportation systems can sustain quality of life and new growth; and
- The potential impacts of global warming.

The unique aspect of PLANYC is its focus on sustainability. Other U.S. cities are "going green" but New York's plan is ambitious. The focus of the PLANYC program is to address:

- Affordable housing;
- Traffic congestion;
- Lack of open space;
- The need to maintain the existing water infrastructure;
- Transportation infrastructure;
- Power infrastructure;
- The need to reduce greenhouse emissions;
- Air quality improvement;
- Brownfields reclamation; and
- Improved waterways.

Interesting that brownfields remediation is identified as a key focus area for a city wide sustainability program. Moreover, the plan does address not only brownfields but environmental justice issues, open space, and power plant siting.

Expect to see more cities focus on building sustainability into long term planning (also note the Denver GreenPrint initiative).

Chinese Cities—the Urbanization of China

Where is China really with regard to establishing clear and enforceable environmental regulations regarding brownfield sites and "building green?"

China appears to be a contradiction with regards to sustainable development. On one hand, during recent months we have seen China establish bold goals to be the leader in new industries such as electric vehicles (a three year goal). On the other hand, China has apparently made little progress in building "eco cities."

Cities such as Dongtan and Huangbaiyu were slated to be world class examples of sustainable cities. However, these cities have remained largely on the drawing boards with no real progress.

A recent article by Christina Larson in Yale 360 explores (http://www.e360.yale.edu/content/feature.msp?id=2138) the bold visions for these cities and the reasons why they appear to be going nowhere. At the heart of the reasons for the failure seems to be a huge disconnect between the planners and designers and the reality of Chinese culture and politics. In essence, international architects collide with on the ground realities.

Dongtan near Shanghai was envisioned and hailed as "the world's first eco-city." The plan called for a city of 50,000 people living in energy efficient buildings with recycled waste used for fuel (similar to Masdar) and recycled water. The original plan was for completion in 2010 with an eventual population of 500,000 inhabitants.

By 2009 no real progress has been made.

Today, almost nothing has been built. Interesting to note that local environmentalists and the press criticized the project because the project was on a wetlands and would impact migratory birds. Even the visitor center has been closed.

A project in Huangbaiyu was planned to transform a small village into an energy efficient community. The project included the use of "special" hay and pressed-earth bricks for construction of homes. This was a failure due to cost overruns and the relatively high price of the homes. Again, a classic error in not understanding the local culture and imposing a vision that doesn't connect with the population it is intended to benefit.

The problems resulting in these failed projects (identified by Christina Larson) are outlined below:

- Lack of understanding of the local culture. According to Richard "Tad" Ferris, Washington, D.C., (Holland & Knight) an "aspirational culture" rather than a "compliance culture" which translates into the implementation and oversight of regulations and plans frequently falls short of reality.

- World class expertise was not a guarantee the project would be a success.

- Funding problems (not unique to Chinese cities) and confusion as to who will pay.

- Lack of oversight during construction.

- Greater enthusiasm for better green building codes than for designing new cities.

- Shortage of stakeholder engagement.

Well this sounds surprisingly negative regarding the prospects for sustainable development and green building in China. Actually, just the opposite. To me these experiments demonstrate a willingness to try something new and to seek outside input. Considering the transformation in China underway as it becomes more urban I expect we will see greater experimentation and a move to not only establish green building codes but to build (or rebuild) sustainable cities. One only needs to look at China's investment in mass transit as an indicator that it is committed to tackling urban development issues.

Connected Cities

IT has the ability to transform cities. The Global e-Sustainability Initiative (GeSI) report issued last year (www.gesi.org) outlines how the Information Technology and Communications (ITC) sector can improve energy efficiency, reduce greenhouse gas emissions, and improve logistics. the GeSI report states that ITC could reduce annual manmade global emissions by 15 percent by 2020 and deliver energy efficiency savings to global businesses of over EUR 500 billion [GBP 400 billion/USD 800 billion].

The report illustrates that while ICT's GHG footprint is roughly 2 percent of global emissions and will almost double by 2020, this is countered by the sector's unique ability to monitor and maximize energy efficiency both within and outside of its own sector could cut CO_2 emissions by up to five times this amount. This represents a savings of 7.8 Gigatonnes of carbon dioxide equivalent ($GtCO_2e$) by 2020—greater than the current annual emissions of either the U.S. or China.

Replacing physical products and services with their virtual equivalents (teleworking, videoconferencing, e-paper, and e-commerce equivalents (dematerialisation and substitution) is only one part (6 percent) of the estimated low carbon benefits the ICT sector can deliver. Greater opportunities for emissions savings exist in applying ICT to global infrastructure and industry in the areas of: smart grid, smart logistics, smart motors, and smart homes.

The GeSI report is well worth the read.

Cisco Systems is one company moving to capitalize on this opportunity through its "Connected Cities" initiative. In March 2009, Cisco signed a Memorandum of Understanding with Metropolis. Metropolis is an association of 106 global cities which is crafting a "2030 Vision." Cisco will assist the Metropolis organization in developing a plan for these sustainable cities. The collaboration will demonstrate how future cities can deploy technology to "advance clean public transportation, decrease pollution through intelligent urban development, offer pervasive access to citizen-centric services, and support sustainable economic growth."

Cisco will also work with Metropolis to develop new practices and enhance existing models in urban development and sustainability. This commitment is an outgrowth and evolution of a current program of the "Connected Urban Development Program" which is a public-private partnership (includes cities such as San Francisco, Amsterdam, and Seoul) which develops ICT solutions to help promote sustainable, intelligent urban development practices.

Other aspects of the agreement include:

• "Cisco and Metropolis will develop a longterm vision for sustainable cities enabled by networked technologies;

- Metropolis and Cisco will explore the possibility of establishing regional innovation centers that will anchor implementation actions on different policy areas, associating the selected cities and Cisco to develop new platforms and tools to manage urban services.

- Cisco will also provide strategic advice on the program and support the development of a partnership approach in urban innovation for Metropolis and its business associates, with a specific focus on technology and business partnerships; and

- Cisco will use its collaborative solutions, such as Cisco WebEx and Cisco TelePresence, to facilitate the creation of a collaborative operating model between Metropolis and its member cities, using a Web 2.0 approach."

Green Commercial Buildings as Brand Symbols

Why does a multinational corporation "build green?" In addition to the reduced operating costs over the life of the building, brand value is a strong driver in the decision to achieve LEED certification for many corporations.

One of the best examples is the Bank of America Building built by the New York-based Durst Organization. The Bank of America Building at One Bryant Park is impressive.

Some of the facts:

- Developers—The Durst Organization has been at the forefront of the environmental movement since its founding in 1915 and in 1999 completed 4 Times Square, recognized as the first "green" high-rise office building in the United States.

- Tennant—Bank of America, a founding financial institution sponsor of the United States Green Building Council known for its environmental stewardship, including its commitment to reduce greenhouse emissions by 7 percent by the year 2008

- Architect—Cook & Fox Architects, New York

- "Green Features" Building on The Durst Organization's and Cook+Fox's commitment to environmentally responsible architecture, the Bank of America Tower will incorporate innovative, high-performance environmental technologies to promote the health and productivity of tenants, reduce waste and assure environmental sustainability:

- Higher ceilings and translucent insulating glass in floor-to-ceiling windows permit maximum daylight in interior spaces, optimal views and energy efficiency
- Advanced double-wall technology provides remarkable views in and out of the building, while dissipating the sun's heat
- Pioneering filtered under-floor displacement air ventilation system and floor-by-floor air handling units allow for individual floor control and more even, efficient, and healthy heating and cooling
- Carbon dioxide monitors automatically adjust the amount of fresh air when necessary
- Gray-water system captures and reuses all rainwater and wastewater, saving millions of gallons of water annually
- Waterless urinals, low-flow fixtures, etc. decrease the use of precious resources
- Thermal storage system at cellar level produces ice in the evening when electricity rates are lowest to reduce peak daytime demand loads on the city
- Daylight dimming and LED lights reduce electric usage
- Recyclable and renewable building materials were used (steel, blast furnace, drywall)
- Green roofs reduce urban heat island effect
- State-of-the-art onsite 5.1-megawatt co-generation plant provides a clean, efficient power source for the building's energy requirements
- 95 percent air filtration
- Environmental goals
- World's most environmentally responsible high-rise office building, focusing on sustainable sites, water efficiency, indoor environmental quality, and energy and atmosphere
- First high-rise to strive for U.S. Green Building Council's Leadership in Energy & Environmental Design "Platinum" designation
- Reduce energy consumption by a minimum of 50 percent
- Reduce potable water consumption by 50 percent

- Reduce storm water contribution by 95 percent
- Utilize 50 percent recycled material in building construction
- Obtain 50 percent of building material within 500 miles of site.

Energy and Brownfields

With the current focus on energy, including renewables, the connection to brownfields redevelopment is worth highlighting as it is the beginning of a longterm trend.

A working paper by the Northeast-Midwest Institute from 2008 highlights the benefits of brownfields and urban redevelopment (A Working Paper by Evans Paull Northeast-Midwest Institute, Updated, April 2008).

In summary the thesis of the paper is that brownfields and urban redevelopment programs have well-documented benefits of restoring neighborhoods, bringing back jobs, cleaning up abandoned factories, and converting eyesores into assets. Several studies have made the connection between urban/brownfields redevelopment and the avoidance of sprawl-related environmental impacts. The bottom line is that infill projects produce substantial air quality and energy-related benefits.

I like the term "sustainable urban redevelopment" as coined by the institute as it describes development that is green and energy-efficient both "internally within the building envelope and externally, in that there are energy savings by virtue of the project location and its relationship to the city." This captures the full value of the redevelopment project.

While green buildings typically save about 30 percent on energy use within the structure, "compact urban development" saves 20 to 40 percent of vehicle miles traveled (VMT) with corresponding reductions in greenhouse gases (GHGs) as previously noted in this chapter.

Brownfields, as a subset of urban redevelopment, have been shown to have similar VMT-related energy benefits. When redevelopment projects combine both elements (VMT reduction and energy-efficient buildings), the energy savings can be estimated to be 30 to 35 percent of the total energy demands attributable to the development, relative to conventional construction in suburban auto-dependent locations.

This is all great news but still doesn't capture full value. According to the institute, the following factors are not accounted for:

- "Urban density is associated with energy efficiencies within the building due to fewer exposed surfaces.

- There is less "line-loss" in distributing electricity to dense urban areas than to spread suburban areas.

- Less energy is spent on building and maintaining infrastructure for urban projects than suburban sprawl projects.

- Some urban projects are served by waste-to-energy plants or district heating systems that also lower GHGs.

- An indirect benefit of urban redevelopment is the retention of greenfield "carbon sinks."

These factors highlight that the promotion of sustainable urban redevelopment can be a major source of greenhouse gas reduction.

Let's take this further. Another interesting aspect of the redevelopment of brownfield sites and energy is the potential reuse of brownfield sites for renewable energy locations.

The USEPA recently examined the rationale for locating renewable energy projects on brownfield sites (http://www.epa.gov/renewableenergyland/why_develop.htm). They include:

- "Many EPA tracked lands, such as large Superfund and RCRA sites, and mining sites offer thousands of acres of land, may be situated in areas where the presence of wind and solar structures are less likely to be met with aesthetic opposition.

- These EPA tracked lands have existing electric transmission lines and capacity and other critical infrastructure, such as roads, and are adequately zoned for such development. The avoided new infrastructure capital and zoning costs is often significant.

- Whether it is a long-term lease or outright purchase, EPA tracked lands may have lower overall transaction costs than greenfields due to the relative ease of acquisition of large swaths of land from one or few owners, versus acquisition of greenfields from potentially numerous landowners.

- Redevelopment of brownfields for "green" energy production can help reduce the stress on greenfields for construction of new energy facilities, and can provide clean, emission-free energy.

- Many EPA tracked lands are in areas where traditional redevelopment may not be an option because the site may be remote, or may simply be saddled with environmental conditions that are not well suited for traditional redevelopment such as residential or commercial.

- Some EPA tracked sites such as industrial, manufacturing, and mining sites were once operations that provided jobs for the local communities. However, once these facilities ceased operations, these same communities were left with fewer jobs. The development, operation, and maintenance of renewable energy facilities on these same sites may reintroduce job opportunities.

- There are approximately 480,000 sites and almost 15 million acres of potentially contaminated properties across the United States that are tracked by EPA. Cleanup goals have been achieved and controls put in place to ensure long-term protection for more than 850,000 acres. This leaves open many potential opportunities to develop renewable energy facilities on these sites, and coordination and partnerships among federal, state, tribal, and other government agencies, utilities, the private sector, and communities, will only help advance renewable energy production."

In view of recent opposition to the location of solar PV installations on open land (desert land in California for example), brownfield sites may represent a way to build renewable energy facilities while addressing the not in my backyard (NIMBY) push back that accompanies even the most beneficial projects such as clean renewable energy.

FINAL THOUGHTS

In closing I wanted to scroll back to a project that was started several decades ago.

Does anyone remember the Arcosanti project in Arizona?

The city was conceived by the Italian-born architect Paolo Soleri (National Design Museum's Lifetime Achievement Award in 2006), as an "urban laboratory" for experiments in sustainable living in 1970. Some people believe this project was the precursor to

Masdar. Arcosanti was an attempt to fuse innovative architecture with the clean technologies available at the time to conserve energy and minimize waste.

The plan was for this to be a demonstration of Soleri's vision for how society could lessen its destructive impact on the environment. According to an interview with Soleri, the "key to making cities instruments of progress rather than models of decline is to integrate all of their systems." This vision and belief has never been more critical and relevant. The project is unfinished and funding for an extension to the site is needed.

Soleri got it right. We need to have a low impact on the environment and capture the sun to power our cities. Essential design principles for a world where we overuse our resources and cities have become "inhuman." We need to return cities to human scale and a more sustainable impact on the environment and our social fabric.

I remain hopeful that the next 20 years will see us move closer to the dreams of OPL, Arcosanti, Masdar, the "post-Kyoto" vision of Paris, and other sustainable cities. Along the way we will clean up brownfield sites and restore them to productive use.

Glossary

The goal of this glossary is to provide the reader with a deeper understanding of the language of brownfields remediation, sustainable land use, green building, and climate change. The glossary is not meant to be complete, but instead provide definitions for key terms used by professionals in the field of sustainable development, and not all of these terms are used in this book.

A

Abandonment—A halt to the use of a property by the owner without the intention of either transferring the rights to the property or resuming use. (www.eli.org)

Adaptation—Adjustment in natural or human systems to a new or changing environment. Adaptation to climate change refers to adjustment in natural or human systems in response to actual or expected climatic stimuli or their effects, which moderates harm or exploits beneficial opportunities. Various types of adaptation can be distinguished, including anticipatory and reactive adaptation, private and public adaptation, and autonomous and planned adaptation. (http://epa.gov/brownfields)

Albedo—The fraction of solar radiation reflected by a surface or object, often expressed as a percentage. Snow covered surfaces have a high albedo; the albedo of soils ranges from high to low; vegetation covered surfaces and oceans have a low albedo. The Earth's albedo varies mainly through varying cloudiness, snow, ice, leaf area, and land cover changes. (http://epa.gov/brownfields)

Alternative Energy—Energy derived from nontraditional sources (e.g., compressed natural gas, solar, hydroelectric, wind). (http://epa.gov/brownfields)

Annex I Countries/Parties—A group of countries included in annex I (as amended in 1998) to the United Nations Framework Convention on Climate Change, including all the developed countries in the Organization of Economic Cooperation and Development, and economies in transition. By default, the other countries are referred to as non-annex I countries. Under Articles 4.2 (a) and 4.2 (b) of the convention, annex I countries commit themselves specifically to the aim of returning individually or jointly to their 1990 levels of greenhouse gas emissions by the year 2000. (http://epa.gov/brownfields)

Anthropogenic—Made by people or resulting from human activities, usually used in the context of emissions that are produced as a result of human activities. (http://epa.gov/brownfields)

Appropriate Technology—Technology that creates minimal environmental impact while serving basic human needs. Uses the simplest level of technology that can effectively achieve the intended purpose in a particular location. (http://www.oikos.com/library/green_building_glossary.html)

Atmosphere—The gaseous envelope surrounding the Earth. The dry atmosphere consists almost entirely of nitrogen (78.1 percent volume mixing ratio) and oxygen (20.9 percent volume mixing ratio), together with a number of trace gases, such as argon (0.93 percent volume mixing ratio), helium, radiatively active greenhouse gases such as carbon dioxide (0.035 percent volume mixing ratio), and ozone. In addition, the atmosphere contains water vapor, whose amount is highly variable but typically 1 percent volume mixing ratio. The atmosphere also contains clouds and aerosols. (http://epa.gov/brownfields)

Atmospheric Lifetime—The lifetime of a greenhouse gas refers to the approximate amount of time it would take for the anthropogenic increment to an atmospheric pollutant concentration to return to its natural level (assuming emissions cease) as a result of either being converted to another chemical compound or being taken out of the atmosphere via a sink. This time depends on the pollutant's sources and sinks as well as its reactivity. The lifetime of a pollutant is often considered in conjunction with the mixing of pollutants in the atmosphere; a long lifetime will allow the pollutant to mix throughout the atmosphere. Average lifetimes can vary from about a week (sulfate aerosols) to more than a century (chlorofluorocarbons [CFCs], carbon dioxide). *See also Greenhouse Gas and Residence Time.* (http://epa.gov/brownfields)

B

Biomass—Total dry weight of all living organisms that can be supported at each tropic level in a food chain. Also, materials that are biological in origin, including organic material (both living and dead) from above and below ground, for example, trees, crops, grasses, tree litter, roots, and animals and animal waste. (http://epa.gov/brownfields)

Biosphere—The part of the Earth system comprising all ecosystems and living organisms, in the atmosphere, on land (terrestrial biosphere) or in the oceans (marine biosphere), including derived dead organic matter, such as litter, soil organic matter, and oceanic detritus. (http://epa.gov/brownfields)

Brownfield—An industrial or commercial property that remains abandoned or underutilized in part because of environmental contamination or the fear of such contamination. (Government definitions of the term may vary depending on the program.) (www.eli.org)

Building Science—Study of how all systems of a structure function together to optimize building performance and prevent building failure. This includes the detailed analysis of energy and moisture flows, building materials, building envelopes, and mechanical systems. (http://www.oikos.com/library/green_building_glossary.html)

C

Cap-And-Trade System—A regulatory or management system that sets a target level for emissions or natural resource use, and, after distributing shares in that quota, lets trading in those permits determine their price. (http://www.oecd.org/glossary)

Carbon Cycle—All parts (reservoirs) and fluxes of carbon. The cycle is usually thought of as four main reservoirs of carbon interconnected by pathways of exchange. The reservoirs are the atmosphere, terrestrial biosphere (usually includes freshwater systems), oceans, and sediments (includes fossil fuels). The annual movements of carbon, the carbon exchanges between reservoirs, occur because of various chemical, physical, geological, and biological processes. The ocean contains the largest pool of carbon near the surface of the Earth, but most of that pool is not involved with rapid exchange with the atmosphere. (http://epa.gov/brownfields)

Carbon Dioxide—A naturally occurring gas, and also a byproduct of burning fossil fuels and biomass, as well as land-use changes and other industrial processes. It is the principal anthropogenic greenhouse gas that affects the Earth's radiative balance. It is the reference gas against which other greenhouse gases are measured and therefore has a global warming potential of 1. *See also Climate Change and Global Warming*. (http://epa.gov/brownfields)

Carbon Dioxide Equivalent—A metric measure used to compare the emissions from various greenhouse gases based upon their global warming potential (GWP). Carbon dioxide equivalents are commonly expressed as "million metric tons of carbon dioxide equivalents ($MMTCO_2Eq$)." The carbon dioxide equivalent for a gas is derived by multiplying the tons of the gas by the associated GWP. The use of carbon equivalents (MMTCE) is declining. (http://epa.gov/brownfields)

$MMTCO_2Eq$ = (million metric tons of a gas) (GWP of the gas)

Carbon Intensity—The amount of carbon by weight emitted per unit of energy consumed. A common measure of carbon intensity is weight of carbon per British thermal unit (Btu) of energy. When there is only one fossil fuel under consideration, the carbon intensity and the emissions coefficient are identical. When there are several fuels, carbon intensity is based on their combined emissions coefficients weighted by their energy consumption levels. (http://epa.gov/brownfields)

Carbon Sequestration—The uptake and storage of carbon. Trees and plants, for example, absorb carbon dioxide, release the oxygen and store the carbon. Fossil fuels were at one time biomass and continue to store the carbon until burned. (http://epa.gov/brownfields)

Carrying Capacity—The maximum population (of humans and other species) that a particular environment can sustain without irreversible environmental damage. (http://www.oecd.org/glossary)

Catchment Area—The area from which rainwater drains into a river, lake, or other body of water. (http://www.oecd.org/glossary)

Certificate of Completion—A written verification from a state voluntary cleanup or brownfield program that a site has been cleaned up in a manner satisfactory to the state. In some states, a certificate provides liability protection, but in most states liability relief must be obtained through another mechanism such as a covenant not to sue. (www.eli.org)

Chlorofluorocarbons—Greenhouse gases covered under the 1987 Montreal Protocol and used for refrigeration, air conditioning, packaging, insulation, solvents, or aerosol propellants. Since they are not destroyed in the lower atmosphere, CFCs drift into the upper atmosphere where, given suitable conditions, they break down ozone. These gases are being replaced by other compounds, including hydrochlorofluorocarbons and hydrofluorocarbons, which are greenhouse gases covered under the Kyoto Protocol. (http://epa.gov/brownfields)

Cleanup Approval Letter—A written verification from a state voluntary cleanup or brownfield program that a site has been cleaned up in a manner satisfactory to the state. (www.eli.org)

Climate—Climate in a narrow sense is usually defined as the "average weather," or more rigorously as the statistical description in terms of the mean and variability of relevant quantities over a period of time ranging from months to thousands of years. The classical period is three decades, as defined by the World Meteorological Organization (WMO). These quantities are most often surface variables such as temperature, precipitation, and wind. Climate in a wider sense is the state, including a statistical description, of the climate system. (http://epa.gov/brownfields)

Climate Change—Climate change refers to any significant change in measures of climate (such as temperature, precipitation, or wind) lasting for an extended period (decades or longer). Climate change may result from:

- Natural factors, such as changes in the sun's intensity or slow changes in the Earth's orbit around the sun;

- Natural processes within the climate system (e.g., changes in ocean circulation);

- Human activities that change the atmosphere's composition (e.g., through burning fossil fuels) and the land surface (e.g., deforestation, reforestation, urbanization, desertification, etc.) (http://epa.gov/brownfields)

Climate Lag—The delay that occurs in climate change as a result of some factor that changes only very slowly. For example, the effects of releasing more carbon dioxide into the atmosphere may not be known for some time because a large fraction is dissolved in the ocean and only released to the atmosphere many years later. (http://epa.gov/brownfields)

Climate Model—A quantitative way of representing the interactions of the atmosphere, oceans, land surface, and ice. Models can range from relatively simple to quite comprehensive. (http://epa.gov/brownfields)

Climate Sensitivity—In IPCC reports, equilibrium climate sensitivity refers to the equilibrium change in global mean surface temperature following a doubling of the atmospheric (equivalent) CO_2 concentration. More generally, equilibrium climate sensitivity refers to the equilibrium change in surface air temperature following a unit change in radiative forcing (degrees Celsius, per watts per square meter, °C/Wm-2). In practice, the evaluation of the equilibrium climate sensitivity requires very long simulations with coupled general circulation models (climate model). The effective climate sensitivity is a related measure that circumvents this requirement. It is evaluated from model output for evolving non-equilibrium conditions. It is a measure of the strengths of the feedbacks at a particular time and may vary with forcing history and climate state. (http://epa.gov/brownfields)

Climate System (or Earth System)—The five physical components (atmosphere, hydrosphere, cryosphere, lithosphere, and biosphere) that are responsible for the climate and its variations. (http://epa.gov/brownfields)

Comfort Letter—A letter issued through a state voluntary cleanup program that typically states that a site complies with the program's requirements, is clean enough for the intended use, and that no future enforcement action is expected unless conditions or uses of the site change. The letter typically does not provide legally enforceable rights such as relief from liability. (www.eli.org)

Community Development Block Grant (CDBG)—A lump-sum grant to a state or local government from the Department of Housing and Urban Development that may be used for development activities including, in some cases, brownfield revitalization. (www.eli.org)

Community Development Corporations (CDCs)—Local nonprofit organizations created to promote urban redevelopment. (www.eli.org)

The Comprehensive Environmental Response, Compensation, and Liability Act (CERCLA or Superfund)—A federal statute that governs the investigation and cleanup of sites contaminated with hazardous substances. The law establishes a trust fund that can be used by the government to clean up sites on the National Priorities List. (www.eli.org)

Concentration—Amount of a chemical in a particular volume or weight of air, water, soil, or other medium. (http://epa.gov/brownfields)

Condemnation—A legal process that allows a government entity to acquire title to property for a public purpose, which, in the case of brownfields, can include removing an unused or potentially contaminated site. Once the property has been condemned, the government entity can destroy any buildings and offer the site for private redevelopment. (www.eli.org)

Conference of the Parties—The supreme body of the United Nations Framework Convention on Climate Change (UNFCCC). It comprises more than 180 nations that have ratified the Convention. Its first session was held in Berlin, Germany, in 1995 and it is expected to continue meeting on a yearly basis. The COP's role is to promote and review the implementation of the Convention. It will periodically review existing commitments in light of the Convention's objective, new scientific findings, and the effectiveness of national climate change programs. (http://epa.gov/brownfields)

Contractor Certification—A process for assuring that contractors meet state standards and have state approval for performing specific tasks. (www.eli.org)

Contractor-Certified Cleanups—Cleanups where the state allows private contractors to make cleanup decisions on behalf of the state, including no-further-action (NFA) letters. Only a small number of states use certified contractors. (www.eli.org)

Contribution Action—A legal proceeding brought by a party that has incurred cleanup costs against other liable parties for their share of the costs incurred. (www.eli.org)

Cooling Degree-Day (CDD)—A way to keep track of the overall cooling demand of a particuluar locaton. CDDs are used to calculate the size of cooling equipment. The value is calculated by adding the average (mean) temperature of each day to a base value, generally 65°F. For example, if the mean temperature over a day was 88, then that day would have 23 cooling degree days. Values are available for hundreds of locations in the United States from the National Oceanic and Atmospheirc Administration's Web page for heating and cooling degree data. (http://www.oikos.com/library/green_building_glossary.html)

Corrective Action—The cleanup process used to address contamination at treatment, storage, and disposal facilities regulated under the Resource Conservation and Recovery Act. (www.eli.org)

Cost-Benefit Analysis—The appraisal of an investment or a policy change that considers all associated costs and benefits, expressed in monetary terms, accruing to it. (http://www.oecd.org/glossary)

Covenant Not to Sue—A written promise by a state government that it will not take legal action or require additional cleanup by a party that satisfactorily cleans up a property under a state brownfield or voluntary cleanup program. (www.eli.org)

D

Decarbonization—When applied to an economy, refers to the phasing out of its dependence on (carbon-containing) fossil fuels. (http://www.oecd.org/glossary)

Deed Restriction—A limitation on the use of a property that is recorded on the deed to the property. The limitations on use are legally enforceable against the owner of the property, but who may enforce the limitation depends on state law. (www.eli.org)

Desertification—The transformation of arid and semi-arid land into desert, generally due to overgrazing, deforestation, poor irrigation and tilling practices, climate change, or a combination of these factors. (http://www.oecd.org/glossary)

Dioxins—A general term that describes a large group of chemicals that are highly persistent in the environment. The most toxic compound is 2,3,7,8-tetrachlorodibenzo-p-dioxin or TCDD. Dioxins are generally formed as unintentional byproducts of industrial processes involving chlorine (such as waste incineration, chemical and pesticide manufacturing, and pulp and paper bleaching), but also during the combustion of biomass, such in wood stoves. (http://www.oecd.org/glossary)

Due Diligence—Evaluation of the environmental condition of a parcel of land, often as part of a real estate transaction. This is required in order for a purchaser to qualify for federal liability protection as an innocent purchaser. (www.eli.org)

E

Easement—A right to use or limit the use of someone else's property. (www.eli.org)

Eco-Label—Information (typically provided on a label attached to a product) informing a potential consumer of a product's characteristics, or of the production or processing method(s) used in its production. (http://www.oecd.org/glossary)

Ecosystem—Any natural unit or entity including living and nonliving parts that interact to produce a stable system through cyclic exchange of materials. (http://epa.gov/brownfields)

Ecosystem Service—A service provided by a group of organisms (including humans in some cases) that is directly or indirectly beneficial to humans. Examples include the conversion of carbon dioxide to oxygen by photosynthesising plants, and the detoxification of harmful chemicals by aquatic and soil-based microbes. (http://www.oecd.org/glossary)

Embodied Energy—The sum total of the energy necessary—from raw material extraction, transport, manufacturing, assembly, installation plus the capital, environmental, and other costs—used to produce a service or product from its beginning through its disassembly, deconstruction, and/or decompostion. (http://www.oikos.com/library/green_building_glossary.html)

Emissions—The release of a substance (usually a gas when referring to the subject of climate change) into the atmosphere. (http://epa.gov/brownfields)

Emissions Factor—A unique value for scaling emissions to activity data in terms of a standard rate of emissions per unit of activity (e.g., grams of carbon dioxide emitted per barrel of fossil fuel consumed). (http://epa.gov/brownfields)

End-of-Pipe Technology—A technology designed to control pollution from another technology, generally installed at the point of emission. (http://www.oecd.org/glossary)

Energy Intensity—The ratio of energy consumption to a measure of the demand for services (e.g., number of buildings, total floor space, floor space-hours, number of employees, or constant dollar value of gross domestic product for services). (http://epa.gov/brownfields)

Engineering Controls—Physical mechanisms for preventing exposure to contamination. Examples include: fences, pavement, and clay caps placed on contaminated soil. (www.eli.org)

Enhanced Greenhouse Effect—The concept that the natural greenhouse effect has been enhanced by anthropogenic emissions of greenhouse gases. Increased concentrations of carbon dioxide, methane, and nitrous oxide, chlorofluorocarbons (CFCs), hydrochlorofluorocarbons (HFCs), perfluorocarbons (PFCs), sulfur hexafluoride (SF_6), nitrogen trifluoride (NF_3), and other photochemically important gases caused by human activities such as fossil fuel consumption, trap more infra-red radiation, thereby exerting a warming influence on the climate. (http://epa.gov/brownfields)

Environment—The ecosystem in which organisms or a species lives, including both the physical environment and the other organisms with which it comes in contact. (http://www.oecd.org/glossary)

Environmental Assessment—A site evaluation or investigation conducted for purposes of determining the extent, if any, of contamination on a property. An assessment can be informal or formal, and can consist of several stages. For example, a phase I assessment, or basic study of possible contamination at a site, is limited to collecting information about past and present site use and inspecting present conditions. A phase II assessment can follow up a phase I assessment with sampling and analysis of suspected contaminated areas of a site. A phase III assessment can either follow up a phase II assessment by gathering information on the exact extent of the contamination or by preparing plans and alternatives for site cleanup. (www.eli.org)

Environmental Insurance—Used to eliminate or reduce the financial risk of a brownfield transaction. In exchange for payment, an insurance company agrees to accept the risk of the owner being held liable under state or federal laws for cleanup costs or damages above a specified amount. (www.eli.org)

Environmental Tax—A tax that is of major relevance for the environment, regardless of its specific purpose or name. (http://www.oecd.org/glossary)

Eutrophication—The process by which a body of water accumulates nutrients, particularly nitrates and phosphates. This process can be accelerated by nutrient-rich runoff or seepage from agricultural land or from sewage outfalls, leading to rapid and excessive growth of algae and aquatic plants and undesirable changes in water quality. (http://www.oecd.org/glossary)

Evapotranspiration—The combined process of evaporation from the Earth's surface and transpiration from vegetation. (http://epa.gov/brownfields)

Exaction—A local government may require an exaction to require concessions from developers, such as the construction of sidewalks on land that will be developed. The exaction must further a legitimate public interest. (www.eli.org)

Externality—A non-market effect on the utility of an individual, or on the costs of a firm, from variables that are under the control of some other agent. (http://www.oecd.org/glossary)

F

Feedback Mechanisms—Factors which increase or amplify (positive feedback) or decrease (negative feedback) the rate of a process. An example of positive climatic feedback is the ice-albedo feedback. (http://epa.gov/brownfields)

Fluorocarbons—Carbon-fluorine compounds that often contain other elements such as hydrogen, chlorine, or bromine. Common fluorocarbons include chlorofluorocarbons (CFCs), hydrochlorofluorocarbons (HCFCs), hydrofluorocarbons (HFCs), and perfluorocarbons (PFCs). *See also Chlorofluorocarbons, Hydrochlorofluorocarbons, Hydrofluorocarbons, Perfluorocarbons, Ozone Depleting Substance.* (http://epa.gov/brownfields)

Footprint (Ecological)—A measure of the hectares of a biologically productive area required to support a human population of given size. (http://www.oecd.org/glossary)

Forcing Mechanism—A process that alters the energy balance of the climate system, i.e., changes the relative balance between incoming solar radiation and outgoing infrared radiation from Earth. Such mechanisms include changes in solar irradiance, volcanic eruptions, and enhancement of the natural greenhouse effect by emissions of greenhouse gases. (http://epa.gov/brownfields)

Foreclosure—A legal action taken by a lender to take the collateral (e.g., a property) that secures the loan and to extinguish the rights of the borrower in the collateral. (www.eli.org)

G

Genetic Diversity—The variation in the genetic composition of individuals within or among species; the heritable genetic variation within and among populations. (http://www.oecd.org/glossary)

Genuine Saving—A measure of sustainable development that corrects the traditional measure of gross savings for the monetary value of the degradation of natural capital, and of the accumulation of human capital. (http://www.oecd.org/glossary)

Geosphere—The soils, sediments, and rock layers of the Earth's crust, both continental and beneath the ocean floors. (http://epa.gov/brownfields)

Geothermal Energy—Literally, the heat of the earth. Where this heat occurs close to the Earth's surface, and is able to maintain a temperature in the surrounding rock or water at or above 150°C, it may be tapped to drive steam turbines. (http://www.oecd.org/glossary)

Global Warming—Global warming is an average increase in the temperature of the atmosphere near the Earth's surface and in the troposphere, which can contribute to changes in global climate patterns. Global warming can occur from a variety of causes, both natural and human induced. In common usage, "global warming" often refers to the warming that can occur as a result of increased emissions of greenhouse gases from human activities. (http://epa.gov/brownfields)

Global Warming Potential (GWP)—Global warming potential (GWP) is defined as the cumulative radiative forcing effects of a gas over a specified time horizon resulting from the emission of a unit mass of gas relative to a reference gas. The GWP-weighted emissions of direct greenhouse gases in the U.S. inventory are presented in terms of equivalent emissions of carbon dioxide (CO_2), using units of teragrams of carbon dioxide equivalents (Tg CO_2 Eq.).

Conversion: Tg = 109 kg = 106 metric tons = 1 million metric tons

The molecular weight of carbon is 12, and the molecular weight of oxygen is 16; therefore, the molecular weight of CO_2 is 44 (i.e., 12+[16 x 2]), as compared to 12 for carbon alone. Thus, carbon comprises $\frac{12}{44}$ of carbon dioxide by weight. (http://epa.gov/brownfields)

Governance—The way that a corporation or government organizes and carries out its economic, political, and administrative authority. (http://www.oecd.org/glossary)

Grandfathering—Granting an existing firm a legal exemption from a new or changed policy. In the case of tradable permits, it refers to the common practice of allocating permits to existing polluters or users of natural resources at no direct cost to them. (http://www.oecd.org/glossary)

Green Building—Construction that increases the efficiency with which buildings use resources—energy, water, and materials—

while reducing building impacts on human health and the environment. May be accomplished by applying these requirements to siting, design, construction, operation, maintenance, and removal—encompassing the entire building life cycle. (http://www.oikos.com/library/green_building_glossary.html)

Greenfield—A property that has not been previously developed. (www.eli.org)

Greenhouse Effect—Trapping and buildup of heat in the atmosphere (troposphere) near the Earth's surface. Some of the heat flowing back toward space from the Earth's surface is absorbed by water vapor, carbon dioxide, ozone, and several other gases in the atmosphere and then reradiated back toward the Earth's surface. If the atmospheric concentrations of these greenhouse gases rise, the average temperature of the lower atmosphere will gradually increase. (http://epa.gov/brownfields)

Greenhouse Gas (GHG)—Any gas that absorbs infrared radiation in the atmosphere. Greenhouse gases include, but are not limited to, water vapor, carbon dioxide (CO_2), methane (CH_4), nitrous oxide (N_2O), chlorofluorocarbons (CFCs), hydrochlorofluorocarbons (HCFCs), ozone (O_3), hydrofluorocarbons (HFCs), perfluorocarbons (PFCs), and sulfur hexafluoride (SF_6). (http://epa.gov/browfields)

H

Habitat—The place or type of site where an organism or population occurs naturally. (http://www.oecd.org/glossary)

Hard Costs—A term used in development projects for the amount that includes total land costs, site clearance, grading and construction costs, and landscaping. (www.eli.org)

Heating Degree-Day (HDD)—A way to keep track of the overall heating demand of a particuluar locaton. HDDs are used to calculate the size of heating equipment and estimate heating costs. The value is calculated by subtracting the average (mean) temperature of each day from a base value, generally 65°F. For example, if the mean temperature over a day was 48, then that day would have 17 heating degree days. The base was selected because it was assumed that homes would need supplemental heat when the temperature dipped below 65°F. Many energy-efficient homes do not need supplemental heat until the temperature drops much lower. However, the standard definition is still used to calculate heating demand. HDD values are available for thousands of locations in the United States from the National Oceanic and Atmospheric Administration's Web page for heating and cooling degree data. (http://www.oikos.com/library/green_building_glossary.html)

Heavy Metal—A high-atomic-weight metal such as arsenic, cadmium, chromium, cobalt, lead, mercury, uranium, or zinc. Heavy metals can be toxic to plants or animals in relatively low concentrations and tend to accumulate in living tissue. (http://www.oecd.org/glossary)

Hot Spots—Specific areas where the level of contamination is very high. (www.eli.org)

Human Capital—The knowledge, skills, competence and attributes embodied in individuals that facilitate the attainment of personal well-being. (http://www.oecd.org/glossary)

Hydrochlorofluorocarbons (HCFCs)—Compounds containing hydrogen, fluorine, chlorine, and carbon atoms. Although ozone depleting substances, they are less potent at destroying stratospheric ozone than chlorofluorocarbons (CFCs). They have been introduced as temporary replacements for CFCs and are also greenhouse gases. (http://epa.gov/brownfields)

Hydrofluorocarbons (HFCs)—Compounds containing only hydrogen, fluorine, and carbon atoms. They were introduced as alternatives to ozone depleting substances in serving many industrial, commercial, and personal needs. HFCs are emitted as by-products of industrial processes and are also used in manufacturing. They do not significantly deplete the stratospheric ozone layer, but they are powerful greenhouse gases with global warming potentials ranging from 140 (HFC-152a) to 11,700 (HFC-23). (http://epa.gov/brownfields)

Hydrologic Cycle—The process of evaporation, vertical and horizontal transport of vapor, condensation, precipitation, and the flow of water from continents to oceans. It is a major factor in determining climate through its influence on surface vegetation, the clouds, snow and ice, and soil moisture. The hydrologic cycle is responsible for 25 to 30 percent of the mid-latitudes' heat transport from the equatorial to polar regions. (http://epa.gov/brownfields)

Hydrosphere—The component of the climate system comprising liquid surface and subterranean water, such as: oceans, seas, rivers, fresh water lakes, underground water, etc. (http://epa.gov/brownfields)

I

Indemnification—An agreement that provides for one party to bear the costs, either directly or by reimbursement, for damages or losses incurred by a second party. (www.eli.org)

Indicator—A summary measure that provides information on the state of, or change in, a system. (http://www.oecd.org/glossary)

Indoor Air Quality—Refers to the content of interior air that could affect health and comfort of building occupants. IAQ may be compromised by microbial contaminants (mold, bacteria), chemicals (outgassing from building materials and finishes, carbon monoxide, radon), allergens, or humidity levels that are too high or too low. (http://www.oikos.com/library/green_building_glossary.html)

Infill Development—Development on vacant or underused sites in a developed area. (www.eli.org)

Infrared Radiation—Radiation emitted by the Earth's surface, the atmosphere, and the clouds. It is also known as terrestrial or long-wave radiation. Infrared radiation has a distinctive range of wavelengths ("spectrum") longer than the wavelength of the red color in the visible part of the spectrum. The spectrum of infrared radiation is practically distinct from that of solar or short-wave radiation because of the difference in temperature between the Sun and the Earth-atmosphere system. (http://epa.gov/brownfields)

Infrastructure—The roads, utility lines, and other public amenities that support property use. (www.eli.org)

Institutional Controls—Legal and administrative mechanisms designed to reduce exposure to contamination. Examples include: deed restrictions, easements, warning signs and notices, and zoning restrictions. (www.eli.org)

Intergovernmental Panel on Climate Change (IPCC)—The IPCC was established jointly by the United Nations Environment Programme and the World Meteorological Organization in 1988. The purpose of the IPCC is to assess information in the scientific and technical literature related to all significant components of the issue of climate change. The IPCC draws upon hundreds of the world's expert scientists as authors and thousands as expert reviewers. Leading experts on climate change and environmental, social, and economic sciences from some 60 nations have helped the IPCC to prepare periodic assessments of the scientific underpinnings for understanding global climate change and its consequences. With its capacity for reporting on climate change, its consequences, and the viability of adaptation and mitigation measures, the IPCC is also looked to as the official advisory body to the world's governments on the state of the science of the climate change issue. For example, the IPCC organized the development of internationally accepted methods for conducting national greenhouse gas emission inventories. (http://epa.gov/brownfields)

Invasive Species—An introduced species that invades natural habitats. (http://www.oecd.org/glossary)

L

Landfill—A land waste disposal site in which waste is generally spread in thin layers, compacted, and covered with a fresh layer of soil each day. (http://epa.gov/brownfields)

LEED—Leadership in Energy and Environmental Design is a green building rating system developed by the U.S. Green Building Council that promotes a whole-building approach to sustainability. It recognizes performance in five key areas of human and environmental health: sustainable site development, water savings, energy efficiency, materials selection and indoor environmental quality. (http://www.oikos.com/library/green_building_glossary.html)

Liability Relief or Liability Release—Protection from liability for contamination provided by a state government as an incentive for brownfield cleanups. Releases vary in scope and form, and can include covenants not to sue and some types of no-further-action letters and certificates of completion. (www.eli.org)

Life Cycle Assessment (LCA)—Assesses the environmental performance of a product or building over its life cycle. This includes raw material extraction, manufacturing, transportation, use, recycling, and disposal. Green Seal is a well known nonprofit organization that utilizes life-cycle analysis to evaluate and certify products and services that have a lesser impact on the environment and human health. (http://www.oikos.com/library/green_building_glossary.html)

M

Man-Made Capital—The manufactured means of production, such as machinery, equipment, and structures, but also non-production related infrastructure, non-tangible assets, and the financial assets that provide command over current and future output streams. Also referred to as "human-made" or "manufactured" capital. (http://www.oecd.org/glossary)

Market Failure—A situation wherein market prices do not reflect the social opportunity cost of production or consumption. External effects or externalities are evidence of a market failure. (http://www.oecd.org/glossary)

Market Price Support—An indicator of the annual monetary value of gross transfers from consumers and taxpayers (where export subsidies are given) to producers (if the difference is positive) or from producers to consumers (if negative) arising from policy measures that create a gap between domestic market prices and the border price of the good or service in question. (http://www.oecd.org/glossary)

Maximum Sustainable Yield—The maximum amount of a renewable resource that can be harvested over an indefinite period without causing its stock to be depleted. (http://www.oecd.org/glossary)

Methane (CH_4)—A hydrocarbon that is a greenhouse gas with a global warming potential most recently estimated at 23 times that of carbon dioxide (CO_2). Methane is produced through anaerobic (without oxygen) decomposition of waste in landfills, animal digestion, decomposition of animal wastes, production and distribution of natural gas and petroleum, coal production, and incomplete fossil fuel combustion. The GWP is from the IPCC's Third Assessment

Report (TAR). For more information visit EPA's methane site. (http://epa.gov/brownfields)

Metric Ton—Common international measurement for the quantity of greenhouse gas emissions. A metric ton is equal to 2205 lbs or 1.1 short tons. *See also Short Ton.* (http://epa.gov/brownfields)

Municipal Solid Waste (MSW)—Residential solid waste and some non-hazardous commercial, institutional, and industrial wastes. This material is generally sent to municipal landfills for disposal. (http://epa.gov/brownfields)

N

National Accounts—The framework for recording the economic transactions of a country in monetary terms. (http://www.oecd.org/glossary)

National Priorities List (NPL)—The Environmental Protection Agency's list of the most serious uncontrolled or abandoned hazardous waste sites. (www.eli.org)

Natural Building—Uses a range of techniques, building systems, and materials that place major emphasis on sustainability. Focus is on durability and the use of locally available, minimally-processed, renewable, recycled, or salvaged resources, as well as those which produce healthy living environments and maintain indoor air quality. Natural building tends to rely on human labor more than technology. (http://www.oikos.com/library/green_building_glossary.html)

Natural Capital—The renewable and non-renewable resources that enter the production process and satisfy consumption needs, as well as environmental assets that have amenity and productive use, and natural features, such as the ozone layer, that are essential for supporting life. (http://www.oecd.org/glossary)

Natural Gas—Underground deposits of gases consisting of 50 to 90 percent methane (CH_4) and small amounts of heavier gaseous hydrocarbon compounds such as propane (C_3H_8) and butane (C_4H_{10}). (http://epa.gov/brownfields)

Natural Resource Damages—Injuries caused to natural resources such as streams, wildlife, and wetlands by contamination from a site. The government can in some cases compel parties responsible for the injuries to pay damages. (www.eli.org)

Net metering—An electricity policy for consumers who own small, renewable energy facilities, such as wind or solar power, or use vehicle-to-grid systems. Net refers to what remains after deductions —the deduction of any energy outflows from metered energy inflows. The customer receives retail credit for the excess electricity generated. (http://www.oikos.com/library/green_building_glossary.html)

Net-Zero-Energy Homes—Homes that create at least as much energy as they use over the course of a typical year. Energy consumption is reduced to very low levels through the use of highly insulated walls, ceilings and floors along with very efficient windows. With space and water heating consumption at very low levels, it is possible to generate enough energy through onsite, renewable energy systems to balance the energy consumed with the energy produced. Currently, solar water heaters and photovoltaic systems are the most common methods of energy generation. This net-zero approach generally means that the building is connected to the electrical grid so that energy can be pulled from the power grid during the winter and pumped into the grid during summer when excess energy is available. (http://www.oikos.com/library/green_building_glossary.html)

Nitrogen Oxides (NO_X)—Gases consisting of one molecule of nitrogen and varying numbers of oxygen molecules. Nitrogen oxides are produced in the emissions of vehicle exhausts and from power stations. In the atmosphere, nitrogen oxides can contribute to the formation of photochemical ozone (smog), can impair visibility, and have health consequences; they are thus considered pollutants. (http://epa.gov/brownfields)

Nitrous Oxide (N_2O)—A powerful greenhouse gas with a global warming potential of 296 times that of carbon dioxide (CO_2). Major sources of nitrous oxide include soil cultivation practices, especially the use of commercial and organic fertilizers, fossil fuel combustion, nitric acid production, and biomass burning. The GWP is from the IPCC's Third Assessment Report (TAR). (http://epa.gov/brownfields)

No-Further-Action (NFA) Letter—A written statement by a state government that it has no present intention to take legal action or require additional cleanup by a party that satisfactorily cleans up a property under a state brownfield or voluntary cleanup program. (www.eli.org)

Non-Market Value—The value of an asset not reflected in market prices. Generally it includes non-use values and those indirect use values (such as certain ecosystem services) and option or quasi-option values for which there is no market. (http://www.oecd.org/glossary)

Non-Methane Volatile Organic Compounds (NMVOCs)—Organic compounds, other than methane, that participate in atmospheric photochemical reactions. (http://epa.gov/brownfields)

Non-Renewable Resource—A resource with a more or less finite initial endowment that can be depleted over time. (http://www.oecd.org/glossary)

Non-Residential Use Standard—A cleanup standard, usually expressed as a numerical ratio of parts of a specific contaminant to parts of the medium of concern (e.g., five parts of lead per million parts of soil) that describes the maximum concentration of the contaminant in the medium that will not present an unacceptable risk to the health of humans engaging in any activity other than residential or those other activities considered to be substantially similar to residential. The non-residential use standard is usually a less strict cleanup standard than the residential use standard, and a site that meets the non-residential standard is limited in its uses to non-residential activities. (www.eli.org)

Non-Use Value—The value to humans derived purely from the fact that an environmental or cultural asset exists, even if they never intend to use it or see it in person. It is can be further subdivided into existence value and bequest value. (http://www.oecd.org/glossary)

O

Off-Grid Electricity—Electricity produced by small generating units that are not connected to high-voltage transmission lines. (http://www.oecd.org/glossary)

Off-the-Grid—Refers to living in a self-sufficient manner without reliance on one or more public utilities. Usually involves a system of generating power that doesn't require connection to utility electricity grids. (http://www.oikos.com/library/green_building_glossary.html)

Ozone (O_3)—Ozone, the triatomic form of oxygen (O_3), is a gaseous atmospheric constituent. In the troposphere, it is created both naturally and by photochemical reactions involving gases resulting from human activities (photochemical smog). In high concentrations, tropospheric ozone can be harmful to a wide range of living organisms. Tropospheric ozone acts as a greenhouse gas. In the stratosphere, ozone is created by the interaction between solar ultraviolet radiation and molecular oxygen (O_2). Stratospheric ozone plays a decisive role in the stratospheric radiative balance. Depletion of stratospheric ozone, due to chemical reactions that may be enhanced by climate change, results in an increased ground-level flux of ultraviolet (UV-) B radiation. (http://epa.gov/brownfields)

Ozone Depleting Substance (ODS)—A family of man-made compounds that includes, but are not limited to, chlorofluorocarbons (CFCs), bromofluorocarbons (halons), methyl chloroform, carbon tetrachloride, methyl bromide, and hydrochlorofluorocarbons (HCFCs). These compounds have been shown to deplete stratospheric ozone, and therefore are typically referred to as ODSs. (http://epa.gov/brownfields)

Ozone Layer—The layer of ozone that begins approximately 15 km above Earth and thins to an almost negligible amount at about 50 km, and shields the Earth from harmful ultraviolet radiation from the sun. The highest natural concentration of ozone (approximately 10 parts per million by volume) occurs in the stratosphere at approximately 25 km above Earth. The stratospheric ozone concentration changes throughout the year as stratospheric circulation changes with the seasons. Natural events such as volcanoes and solar flares can produce changes in ozone concentration, but man-made changes are of the greatest concern. (http://epa.gov/brownfields)

P

Particulate Matter (PM)—Very small pieces of solid or liquid matter such as particles of soot, dust, fumes, mists, or aerosols. The physical characteristics of particles, and how they combine with other particles, are part of the feedback mechanisms of the atmosphere. (http://epa.gov/brownfields)

Parts Per Billion (ppb)—The number of parts of a chemical found in one billion parts of a particular gas, liquid, or solid mixture. (http://epa.gov/brownfields)

Parts Per Million (ppm)—The number of parts of a chemical found in one million parts of a particular gas, liquid, or solid. (http://epa.gov/brownfields)

Passive Solar—Means of using sunlight for energy without active mechanical systems. Converts sunlight into usable heat (water, air, thermal mass), causes air-movement for ventilating, and stores heat for future use without the assistance of other energy sources. Passive solar systems have little to no operating costs, often have low maintenance costs, and emit no greenhouse gases in operation. Requires careful site planning, selection of building materials, and building features. Energy conservation reduces the needed size of any renewable or conventional energy system, and greatly enhances the economics, so it must be performed first. (http://www.oikos.com/library/green_building_glossary.html)

Perfluorocarbons (PFCs)—A group of human-made chemicals composed of carbon and fluorine only. These chemicals (predominantly CF_4 and C_2F_6) were introduced as alternatives, along with hydrofluorocarbons, to the ozone depleting substances. In addition, PFCs are emitted as byproducts of industrial processes and are also used in manufacturing. PFCs do not harm the stratospheric ozone layer, but they are powerful greenhouse gases: CF_4 has a global warming potential (GWP) of 5700 and C_2F_6 has a GWP of 11,900. The GWP is from the IPCC's Third Assessment Report (TAR). (http://epa.gov/brownfields)

Persistent Organic Pollutant—A complex organic chemical which resists decomposition in the environment and can migrate over great distances, which bioaccumulates and biomagnifies, and which is suspected of being toxic to humans or other organisms exposed to even low concentrations if such exposure occurs over a long period of time. Examples include certain pesticides (aldrin, chlordane, DDT, dieldrin, endrin, heptachlor, mirex, and toxaphene), industrial chemicals (PCBs and hexachlorobenzene, which is also a pesticide), and unwanted byproducts of combustion and industrial processes (dioxins and furans). (http://www.oecd.org/glossary)

Photosynthesis—The process by which plants take CO_2 from the air (or bicarbonate in water) to build carbohydrates, releasing O_2 in the process. There are several pathways of photosynthesis with different responses to atmospheric CO_2 concentrations. (http://epa.gov/brownfields)

Photovoltaic (Solar) Cell—Generally speaking, a device incorporating a semiconductor that generates electricity when exposed to (sun)light. The technology may be further sub-divided into crystalline, multi-crystalline, thin-film, and concentrator variants. (http://www.oecd.org/glossary)

Polluter Pays Principle—The principle that a polluter should bear the expenses of carrying out pollution prevention and control measures decided by public authorities, to ensure that the environment is in an acceptable state (i.e., costs of these measures should be reflected in the cost of goods and services which cause pollution). (http://www.oecd.org/glossary)

Potentially Responsible Party (PRP)—Under the Comprehensive Environmental Response, Compensation, and Liability Act (CERCLA), a party potentially liable for cleanup costs at a Superfund site. (www.eli.org)

Precaution—Action taken in the face of unresolved uncertainty, especially if the costs of inaction are potentially both high and irreversible. (http://www.oecd.org/glossary)

Pressure-State-Response—A framework for the presentation of environmental information in terms of indicators of the pressures that human activities exert on the environment, of the state of the environment, and of society's responses. (http://www.oecd.org/glossary)

Pro Forma—Financial projections for a real estate project, which include an income statement showing capital costs, operating revenues and expenses, and return on investment over a period of time. (www.eli.org)

Producer Support Estimate—An indicator of the annual monetary value of gross transfers from consumers and taxpayers to producers (measured at the producer's property), arising from policy measures, regardless of their nature, objectives, or impacts on production or income. The percentage PSE is the ratio of the PSE to the value of total gross farm receipts, measured by the value of total production (at farm gate prices), plus budgetary support. (http://www.oecd.org/glossary)

Prospective Purchaser Agreement—An agreement between the Environmental Protection Agency (EPA) and the prospective buyer of a Superfund site that protects the prospective buyer from certain liabilities for contamination that is already on the site, usually in exchange for a payment of money and other commitments by the prospective purchaser. States may also have similar agreements as part of their voluntary cleanup or brownfield programs. (www.eli.org)

Public Good—A special kind of externality in consumption where the availability of a good to one individual does not reduce its availability to others (non-rivalry) and the supplier of the good cannot exclude anybody from consuming it (non-excludability). (http://www.oecd.org/glossary)

R

Radiation—Energy transfer in the form of electromagnetic waves or particles that release energy when absorbed by an object. (http://epa.gov/brownfields)

Radiative Forcing—Radiative forcing is the change in the net vertical irradiance (expressed in Watts per square metre: Wm-2) at the tropopause due to an internal change or a change in the external forcing of the climate system, such as, for example, a change in the concentration of carbon dioxide or the output of the Sun. Usually radiative forcing is computed after allowing for stratospheric temperatures to readjust to radiative equilibrium, but with all tropospheric properties held fixed at their unperturbed values. Radiative forcing is called instantaneous if no change in stratospheric temperature is accounted for. Practical problems with this definition, in particular with respect to radiative forcing associated with changes, by aerosols, of the precipitation formation by clouds, are discussed in Chapter 6 of the IPCC Third Assessment Report Working Group I: The Scientific Basis. (http://epa.gov/brownfields)

Recycling—Collecting and reprocessing a resource so it can be used again. An example is collecting aluminum cans, melting them down, and using the aluminum to make new cans or other aluminum products. (http://epa.gov/brownfields)

Reforestation—Planting of forests on lands that have previously contained forests but that have been converted to some other use. (http://epa.gov/brownfields)

Regulatory Capture—The capacity of narrow interest groups to shape regulations to suit their own goals. (http://www.oecd.org/glossary)

Regulatory Impact Analysis—The ex-ante analysis of the effects of a proposed regulation, or the ex-post assessment of an existing one. (http://www.oecd.org/glossary)

Renewable Resources—Natural resources qualify as renewable resources if they are replenished by natural processes at a rate comparable or faster than their rate of consumption by users (solar radiation, ocean tides, wind). Renewable resources may also include commodities such as wood, bamboo, and crop waste. (http://www.oikos.com/library/green_building_glossary.html)

Representations and Warranties—Statements of fact (representations) and promises (warranties) that a seller makes to a buyer in a real estate transaction. (www.eli.org)

Request-for-Proposals (RFPs)—A document that asks developers for a detailed proposal on development of a site. Proposals may include discussion of the developer's experience and qualifications and project-specific information on market feasibility, urban design, architecture, community appropriateness, and projected financial performance. (www.eli.org)

Reserve—In geology, a reserve refers to an estimated quantity of a natural material (mineral, mineraloid, rock, gas, or liquid) in the ground that has been explored to the extent that the probability of producing the material from it economically (at current market prices and with available technology) is reasonably assured. Reserves are sub-sets of, and not synonymous with, resources. (http://www.oecd.org/glossary)

Residence Time—The average time spent in a reservoir by an individual atom or molecule. With respect to greenhouse gases, residence time usually refers to how long a particular molecule remains in the atmosphere. (http://epa.gov/brownfields)

Residential Use Standard—A cleanup standard, usually expressed as a numerical ratio of parts of a specific contaminant to parts of the medium of concern (e.g., five parts of lead per million parts of soil) that describes the maximum concentration of the contaminant in the medium that will not present an unacceptable risk to the health of humans residing on the site, or engaging in activities on the site that are considered to be substantially similar to residing on the site. The residential use standard is usually the strictest cleanup standard, and a site that meets this standard can usually be used for any purpose. (www.eli.org)

Resource—Generally, a tangible asset. In geology, resources refer to accumulations of natural materials that are known or expected to exist and for which there is a reasonable assurance that a given quantity of the material can be recovered economically at current or expected future market prices using currently available technologies or technologies that can reasonably be expected to become available in the foreseeable future. (http://www.oecd.org/glossary)

The Resource Conservation and Recovery Act (RCRA)—A federal statute that regulates the generation, transportation, storage, treatment, and disposal of hazardous waste. RCRA programs include the Corrective Action and Underground Storage Tank Programs. (www.eli.org)

Respiration—The process whereby living organisms convert organic matter to CO_2, releasing energy and consuming O_2. (http://epa.gov/brownfields)

Restrictive Covenant—A specific type of deed restriction. For example, a restrictive covenant could prohibit commercial uses. (www.eli.org)

Risk Assessment—A study or evaluation that identifies, and in many cases quantifies, the potential harm posed to health and the environment by contamination on a property. (www.eli.org)

Running With the Land—An obligation or right that attaches to a property and passes to the new owner after the land is sold. (www.eli.org)

S

Short Ton—Common measurement for a ton in the United States. A short ton is equal to 2000 lbs or 0.907 metric tons. (http://epa. gov/brownfields)

Sick Building Syndrome—Often caused by flaws in heating, ventilation, and air conditioning (HVAC) systems. Contaminants produced by outgassing of some types of building materials, volatile organic compounds, molds, improper exhausting of light industrial and cleaning chemicals have also been attributed to SBS. (http://www.oikos.com/library/green_building_glossary.html)

Sink—Any process, activity, or mechanism which removes a greenhouse gas, an aerosol, or a precursor of a greenhouse gas or aerosol from the atmosphere. (http://epa.gov/brownfields)

Social Capital—The networks and shared norms, values, and understanding that facilitate cooperation within and between groups. (http://www.oecd.org/glossary)

Soil Carbon—A major component of the terrestrial biosphere pool in the carbon cycle. The amount of carbon in the soil is a function of the historical vegetative cover and productivity, which in turn is dependent in part upon climatic variables. (http://epa.gov/browfields)

Solar Radiation—Radiation emitted by the Sun. It is also referred to as short-wave radiation. Solar radiation has a distinctive range of wavelengths (spectrum) determined by the temperature of the Sun. (http://epa.gov/brownfields)

Stakeholder—Somebody who has a "stake" or interest in a public policy, program or, in some uses of the term, a corporation's activities. (http://www.oecd.org/glossary)

Stratosphere—Region of the atmosphere between the troposphere and mesosphere, having a lower boundary of approximately 8 km at the poles to 15 km at the equator and an upper boundary of approximately 50 km. Depending upon latitude and season, the temperature in the lower stratosphere can increase, be isothermal, or even decrease with altitude, but the temperature in the upper stratosphere generally increases with height due to absorption of solar radiation by ozone. (http://epa.gov/brownfields)

Streamflow—The volume of water that moves over a designated point over a fixed period of time. It is often expressed as cubic feet per second (ft3/sec). (http://epa.gov/brownfields)

Sulfur Hexafluoride (SF_6)—A colorless gas soluble in alcohol and ether, slightly soluble in water. A very powerful greenhouse gas used primarily in electrical transmission and distribution systems and as a dielectric in electronics. The global warming potential of SF_6 is 22,200. This GWP is from the IPCC's Third Assessment Report (TAR). (http://epa.gov/brownfields)

Superfund—See the Comprehensive Environmental Response, Compensation, and Liability Act. (www.eli.org)

Sustainable Development—Development that meets the needs of the present without compromising the ability to meet the needs of the future. Encompasses three parts: environmental sustainability, economic sustainability, and social-political sustainability. (http://www.oikos.com/library/green_building_glossary.html)

T

Tax Increment Financing (TIF)—A mechanism that allows local governments to use future projected taxes to finance current infrastructure investments. (www.eli.org)

Tax Credit—Incentives to invest in a development that reduce liability for taxes that otherwise would be incurred. (www.eli.org)

Technology Foresight—A process for establishing common views on future technology development strategies. Typically it seeks views from a large number of communities, including civic groups, as well as academic, government, and industrial research bodies. (http://www.oecd.org/glossary)

Thermohaline Circulation—Large-scale density-driven circulation in the ocean, caused by differences in temperature and salinity. In the North Atlantic the thermohaline circulation consists of warm surface water flowing northward and cold deep water flowing southward, resulting in a net poleward transport of heat. The surface water sinks in highly restricted sinking regions located in high latitudes. (http://epa.gov/brownfields)

Threshold—When used in reference to a species, an ecosystem, or another natural system, it refers to the level beyond which further deterioration is likely to precipitate a sudden adverse, and possibly irreversible, change. (http://www.oecd.org/glossary)

Toxic Tort Action—A legal proceeding brought to seek damages for personal injury or property damage incurred as a result of exposure to a hazardous substance. (www.eli.org)

Trace Gas—Any one of the less common gases found in the Earth's atmosphere. Nitrogen, oxygen, and argon make up more than 99 percent of the Earth's atmosphere. Other gases, such as carbon dioxide, water vapor, methane, oxides of nitrogen, ozone, and ammonia, are considered trace gases. Although relatively unimportant in terms of their absolute volume, they have significant effects on the Earth's weather and climate. (http://epa.gov/brownfields)

Troposphere—The region of the atmosphere closest to the Earth, extending from the surface up to about 10 km in altitude (its exact height varies with latitude). Almost all weather takes place in the troposphere. (http://www.oecd.org/glossary)

U

Ultraviolet Radiation (UV)—The energy range just beyond the violet end of the visible spectrum. Although ultraviolet radiation constitutes only about 5 percent of the total energy emitted from the sun, it is the major energy source for the stratosphere and mesosphere, playing a dominant role in both energy balance and chemical composition. Most ultraviolet radiation is blocked by Earth's atmosphere, but some solar ultraviolet penetrates and aids in plant photosynthesis and helps produce vitamin D in humans. Too much ultraviolet radiation can burn the skin, cause skin cancer and cataracts, and damage vegetation. (http://epa.gov/brownfields)

Uncertainty Premium—The amount that the buyer of a property subtracts or discounts from the purchase price to reflect the risk of unexpected environmental assessment and cleanup costs. (www.eli.org)

United Nations Framework Convention on Climate Change (UNFCCC)—The Convention on Climate Change sets an overall framework for intergovernmental efforts to tackle the challenge posed by climate change. It recognizes that the climate system is a shared resource whose stability can be affected by industrial and other emissions of carbon dioxide and other greenhouse gases. The convention enjoys near universal membership, with 189 countries having ratified.

Under the convention, governments:

- Gather and share information on greenhouse gas emissions, national policies, and best practices
- Launch national strategies for addressing greenhouse gas emissions and adapting to expected impacts, including the provision of financial and technological support to developing countries
- Cooperate in preparing for adaptation to the impacts of climate change

The convention entered into force on 21 March 1994. (http://epa.gov/brownfields)

Use Permit—A type of variance that authorizes an otherwise unacceptable use on a property without changing its zoning. (www.eli.org)

Use Value—A value obtained through the use of an environmental or cultural asset. (http://www.oecd.org/glossary)

V

Variance—An individual exception to a land-use restriction or other legal standard granted because of special circumstances. (www.eli.org)

Volatile Organic Compound—A carbon-containing compound, such as gasoline or acetone, that vaporises at a relatively low temperature, generally below 40°C. VOCs can contaminate water, and in the atmosphere can react with other gases in the presence of sunlight to form ozone or other photochemical oxidants. (http://www.oecd.org/glossary)

Voluntary Cleanups—Cleanups of identified contamination that are not court or agency ordered. Most states have voluntary cleanup programs that encourage voluntary cleanups and that may provide benefits if volunteers meet specified standards. (www.eli.org)

W

Wastewater—Water that has been used and contains dissolved or suspended waste materials. (http://epa.gov/brownfields)

Water Vapor—The most abundant greenhouse gas, it is the water present in the atmosphere in gaseous form. Water vapor is an important part of the natural greenhouse effect. While humans are not significantly increasing its concentration, it contributes to the enhanced greenhouse effect because the warming influence of greenhouse gases leads to a positive water vapor feedback. In addition to its role as a natural greenhouse gas, water vapor plays an important role in regulating the temperature of the planet because clouds form when excess water vapor in the atmosphere condenses to form ice and water droplets and precipitation. (http://epa.gov/brownfields)

Weather—Atmospheric condition at any given time or place. It is measured in terms of such things as wind, temperature, humidity, atmospheric pressure, cloudiness, and precipitation. In most places, weather can change from hour-to-hour, day-to-day, and season-to-season. Climate in a narrow sense is usually defined as the "average weather," or more rigorously as the statistical description in terms of the mean and variability of relevant quantities over a period of time ranging from months to thousands or millions of years. The classical period is 30 years, as defined by the World Meteorological Organization (WMO). These quantities are most often surface variables such as temperature, precipitation, and wind. Climate in a wider sense is the state, including a statistical description, of the climate system. A simple way of remembering the difference is that climate is what you expect (e.g., cold winters) and "weather" is what you get (e.g., a blizzard). (http://epa.gov/brownfields)

Willingness-To-Pay—The amount an individual is willing to pay to acquire some good or service. This amount can be elicited from the individual's stated or revealed preferences. (http://www.oecd.org/glossary)

Index

ML   4/10